Research Reactor Aluminum Spent Fuel

Treatment Options for Disposal

Milton Levenson, Principal Investigator
Kevin D. Crowley, Study Director

Board on Radioactive Waste Management
Commission on Geosciences, Environment, and Resources
National Research Council

D0807852

NATIONAL ACADEMY PRESS
Washington, D.C.   1998

NOTICE: The project that is the subject of this report was approved by the Governing Board of the National Research Council, whose members are drawn from the councils of the National Academy of Sciences, the National Academy of Engineering, and the Institute of Medicine.

Support for this study was provided by the U.S. Department of Energy, under Grant No. DE-FC01-94EW54069. All opinions, findings, conclusions, and recommendations expressed herein are those of the authors and do not necessarily reflect the views of the Department of Energy.

Library of Congress Catalog Card No. 98-84670
International Standard Book Number 0-309-06049-4

Additional copies of this report are available from:

National Academy Press
2101 Constitution Avenue, N.W.
Box 285
Washington, DC 20055
800-624-6242
202-334-3313 (in the Washington Metropolitan Area)
http://www.nap.edu

Cover photo: Radiation glow from an empty zirconium oxide crucible that had been used to melt highly radioactive fuel at Argonne West. The glow is from residual fission products on the surface of the crucible which is at room temperature. The ambient light level was 100 footcandles and the photo was taken through a 60-inch thick leaded glass window. Photo courtesy of Argonne National Laboratory.

Printed in the United States of America.

RESEARCH REACTOR ALUMINUM SPENT FUEL
Treatment Options for Disposal

MILTON LEVENSON, Principal Investigator, Menlo Park, California
KEVIN D. CROWLEY, Study Director
ANGELA R. TAYLOR, Senior Project Assistant
LATRICIA BAILEY, Project Assistant

Milton Levenson is a chemical engineer with more than 48 years of experience in nuclear energy and related fields. His technical experience includes work in nuclear safety, fuel cycle, water reactor technology, advanced reactor technology, remote control technology, and sodium reactor technology. His professional experience includes research and operations positions at Oak Ridge National Laboratory, Argonne National Laboratory, the Electric Power Research Institute, and Bechtel Power Corporation. Mr. Levenson is the past president of the American Nuclear Society; a fellow of the American Nuclear Society and the American Institute of Chemical Engineers; and the recipient of the American Institute of Chemical Engineers' Robert E. Wilson Award. He is the author of more than 150 publications and presentations and holds three U.S. patents. He received his B.Ch.E. from the University of Minnesota and was elected to the National Academy of Engineering in 1976.

Kevin D. Crowley is director of the Board on Radioactive Waste Management at the National Research Council and has 15 years of experience in geoscience and nuclear science research and policy work. He holds a Ph.D. degree in geology from Princeton University and previously held positions at the U.S. Geological Survey, the University of Oklahoma, and Miami University of Ohio. He is the author of about 30 technical publications, holds one U.S. patent, and has directed or staffed about a dozen NRC studies.

The National Academy of Sciences is a private, nonprofit, self-perpetuating society of distinguished scholars engaged in scientific and engineering, research, dedicated to the furtherance of science and technology and to their use for the general welfare. Upon the authority of the charter granted to it by the Congress in 1863, the Academy has a mandate that requires it to advise the federal government on scientific and technical matters. Dr. Bruce Alberts is president of the National Academy of Sciences.

The National Academy of Engineering was established in 1964, under the charter of the National Academy of Sciences, as a parallel organization of outstanding engineers. It is autonomous in its administration and in the selection of its members, sharing with the National Academy of Sciences the responsibility for advising the federal government. The National Academy of Engineering also sponsors engineering programs aimed at meeting national needs, encourages education and research, and recognizes the superior achievements of engineers. Dr. William A. Wulf is president of the National Academy of Engineering.

The Institute of Medicine was established in 1970 by the National Academy of Sciences to secure the services of eminent members of appropriate professions in the examination of policy matters pertaining to the health of the public. The Institute acts under the responsibility given to the National Academy of Sciences by its congressional charter to be an adviser to the federal government and, upon its own initiative, to identify issues of medical care, research, and education. Dr. Kenneth Shine is president of the Institute of Medicine.

The National Research Council was organized by the National Academy of Sciences in 1916 to associate the broad community of science and technology with the Academy's purposes of furthering knowledge and of advising the federal government. Functioning in accordance with general policies determined by the Academy, the Council has become the principal operating agency of both the National Academy of Sciences and the National Academy of Engineering in providing services to the government, the public, and the scientific and engineering communities. The Council is administered jointly by both Academies and the Institute of Medicine. Dr. Bruce Alberts and Dr. William A. Wulf are chairman and interim vice-chairman, respectively, of the National Research Council.

FOREWORD

Unlike most National Research Council (NRC) studies, which are undertaken by an appointed committee of experts, this project was conducted by a principal investigator (P.I.), Milton Levenson, a member of the National Academy of Engineering, who was selected for his extensive experience with nuclear fuel cycle, nuclear safety, remote control technology, and reactor technology. The findings presented in this report reflect his views, based on the information made available to him by the study's sponsor. The study was conducted under the aegis of the NRC Board on Radioactive Waste Management.

In preparing this report, the P.I. reviewed data and documents on the aluminum spent fuel program and related issues that were provided to him by the U.S. Department of Energy (Appendix F), and he obtained briefings from DOE and contractor staff at two public information-gathering meetings held near the Savannah River site (Appendix B). In addition, the P.I. obtained technical advice from 13 expert consultants (Appendix E), eleven of whom attended the second information-gathering meeting. These consultants provided the P.I. with short written reports which have been included in Appendix D of this report.

A draft of this report was reviewed by individuals chosen for their diverse perspectives and technical expertise, in accordance with procedures approved by the NRC's Report Review Committee. The purpose of this independent review was to provide candid and critical comments to assist the NRC in making the published report as sound as possible and to ensure that the report meets institutional standards for objectivity, evidence, and responsiveness to the study charge. The content of the review comments and draft manuscript remain confidential to protect the integrity of the deliberative process. We wish to thank the following individuals for their participation in the review of this report.

Patrick R. Atkins, Aluminum Company of America
Donald A. Brand, NAE, Pacific Gas and Electric Company (retired)
Thomas B. Cochran, Natural Resources Defense Council
Harold K. Forsen, NAE, Bechtel Corporation (retired)
B. John Garrick, NAE, PLG, Inc. (retired)

Darleane C. Hoffman, Lawrence Berkeley National Laboratory
William E. Kastenberg, NAE, University of California at Berkeley
Ronald A. Knief, Ogden Environmental and Energy Services Co.
D. Warner North, DFI/Aeronomics
Frank L. Parker, NAE, Vanderbilt University
Steve Pawel, Oak Ridge National Laboratory
Martin J. Steindler, Argonne National Laboratory (retired)
William G. Sutcliffe, Lawrence Livermore National Laboratory
Edwin L. Zebroski, NAE, Aptech Engineering Services, Inc.

Although the individuals listed above provided many constructive comments and suggestions, the responsibility for the final content of this report rests with the P.I. and the NRC.

Bruce Alberts, Chairman
National Research Council

PREFACE

Aluminum spent fuel from foreign and domestic research reactors represents only a small part of the total DOE and commercial spent fuel inventory—less than 10 percent by volume of DOE's inventory and less than 1 percent by volume of the commercial spent fuel inventory. However, aluminum spent fuel represents a challenge for disposal because of its relatively high uranium-235 enrichment. For policy reasons, DOE is seeking alternate options to conventional reprocessing for safe treatment and eventual disposal of this fuel.

DOE chartered a task team to review feasible treatment options and make recommendations, and the Savannah River Office of DOE was assigned the responsibility for implementing them. DOE-Savannah River requested that the National Research Council (NRC) conduct a review of its plans to treat for disposal the aluminum spent research reactor fuel under its management (Appendix A). This report is the product of that review.

Because of the perceived urgency of the aluminum spent fuel program this has been a fast-track review—two months for data collection, one month for writing, and three months for the NRC review process—and has been possible only through the help and cooperation on the part of many people. We involved several expert consultants in our second information-gathering meeting (Appendix B). They adjusted their personal schedules with only one to two weeks advance notice to attend this meeting and provide written reports before the end of the year (these reports appear in Appendix D). The Savannah River staff of DOE and the site contractors, supported by staff from the Yucca Mountain Project, were flexible and responsive to our many requests for information and much more cooperative than has been this writer's past experience with DOE projects of a similar nature. They also have introduced an additional urgency into getting this report published—they are moving rapidly to correct some of the shortcomings identified by the task team report, by their own reviews, and through discussions at the two information-gathering meetings for this project. To be relevant we had to publish quickly.

This report is the result of the Principal Investigator's digestion of a large amount of written information (Appendix F), information provided in the two information-gathering meetings, and a review of the consultants' reports (Appendix D). This was made possible to a large extent by the support and contributions of the NRC staff. Because of the short schedule and the large amount of paper to be reviewed, the work of Angela Taylor and Latricia Bailey in arranging the travel and meetings and turning the paper mill in high gear was essential in meeting our schedule. If this report is coherent, it is due to the considerable writing skills and dedication of Kevin Crowley who took my rambling thoughts and "what about" nuggets and transformed them into regular sentences, paragraphs, and chapters.

This is not a personal report—it has been subjected to the National Research Council's review process and is an NRC report—but if it contains errors of either omission or commission, they are mine.

Milton Levenson

CONTENTS

SUMMARY

The Department of Energy (DOE) Savannah River site, which is located near Aiken, South Carolina, is responsible for managing DOE's inventory of aluminum spent nuclear fuel[1] from foreign and domestic research reactors. During the next four decades, Savannah River will be responsible for receiving and storing approximately 42 MTHM[2] of aluminum spent fuel at the site, processing it as necessary to put it into "road-ready" form[3] for eventual shipment to a repository, and providing for interim storage of the road-ready product until a repository is ready to accept it.

In 1995, the Office of Spent Fuel Management of DOE established the Research Reactor Spent Nuclear Fuel Task Team, or "Task Team," to help develop a technical strategy for disposal of this aluminum spent fuel. The Task Team was asked to evaluate alternative treatment and packaging technologies that could be used in the place of conventional reprocessing[4] to treat for disposal the aluminum spent fuel in a safe and cost-effective manner. The need to develop alternative treatment technologies for aluminum spent fuel was necessitated by

---

[1] Spent nuclear fuel is irradiated fuel that contains uranium-aluminum matrix fuel elements and (or) is clad in aluminum. DOE refers to this fuel as "aluminum-based" or "aluminum-clad" spent fuel, but in this report this fuel will be referred to as "aluminum spent fuel."

[2] Metric tons heavy metal, the amount of heavy metal (uranium, thorium, and plutonium) present in fresh (unirradiated) fuel.

[3] That is, the fuel will be treated and loaded into disposable canisters suitable for interim storage, shipping, and loading at the repository site into a final repository package.

[4] Conventional reprocessing involves the dissolution of aluminum spent fuel in acid followed by the chemical recovery of uranium and plutonium if present.

DOE's policy to phase out reprocessing at the Idaho National Engineering and Environmental Laboratory and Savannah River.[5] The Task Team evaluated eleven treatment technologies and recommended a strategy for selecting, developing, and implementing one or more of these options by the year 2000.

The treatment technologies evaluated by the Task Team fell into one of three categories: (1) direct disposal technologies, which involve no processing of the spent fuel except for drying; (2) highly enriched uranium (HEU)[6] dilution technologies, in which the fuel is physically or metallurgically diluted with depleted uranium[7] to reduce the amount of uranium-235 (^{235}U) to 20 percent or less by mass;[8] and (3) advanced treatment technologies, in which the fuel is processed to produce more acceptable waste forms for repository disposal than is possible for either the direct disposal or the HEU dilution technologies and to reduce the volume of waste for disposal.

The Task Team used a combination of qualitative and quantitative methodologies to screen and rank the alternative treatment options, and it performed a sensitivity analysis of the results. The objective of this exercise was to eliminate from further consideration those options that were less likely to be implemented successfully because of technical, cost, or scheduling difficulties. The Task Team recommended direct co-disposal treatment, which involves the placement

---

[5] The decision to phase-out reprocessing at these sites was announced by the Secretary of Energy in February 1992. This decision was based on recommendations from the Highly Enriched Uranium Task Force, which produced a predecisional draft report on spent fuel reprocessing in February 1992. A copy of this draft report was not made available to the P.I. until the present report was in the final stages of review.

[6] Defined as a material that contains more than 20 percent uranium-235 (^{235}U) by mass. In contrast, natural uranium contains about 0.7 percent ^{235}U by mass.

[7] Depleted uranium is depleted in ^{235}U relative to natural abundances.

[8] The diluted product is known as low-enriched uranium (LEU).

of dried spent fuel into a canister for disposal in a larger package containing high-level waste glass logs (Chapter 2), as the primary treatment option, and melt and dilute treatment, in which the fuel is melted and diluted with depleted uranium, as a "parallel" option. The Task Team also recommended electrometallurgical treatment, which is essentially electrochemical reprocessing, as a backup option, because this technology is fundamentally different than the others and thus offers some protection against unforeseen technical or licensing problems.

As part of its efforts to prepare an environmental impact statement (EIS) for disposing of the aluminum spent nuclear fuel under its management, the Savannah River Office of DOE asked the National Research Council (NRC) to review the alternative treatment options that it has identified to put this fuel into a form suitable for shipment to and emplacement in a repository (Appendix A). The statement of task for this study involved the examination of the following aspects of DOE's program for selecting and implementing a treatment option for aluminum spent fuel:

  • examination of the set of technologies chosen by DOE and identification of other alternatives that DOE might consider;
  • examination of the waste-package performance criteria developed by DOE to meet anticipated waste acceptance criteria for disposal of aluminum spent fuel and identification of other factors that DOE might consider; and
  • to the extent possible given the schedule for this project, an assessment of the cost and timing aspects associated with implementation of each spent nuclear fuel treatment technology.

This study is focused primarily on the treatment step of the disposal process, that is, the options for treating the aluminum spent fuel to make it acceptable for disposal in a repository. This report does not review the other components of the disposal program—for example, the shipment of aluminum spent fuel to Savannah River for storage and treatment, the shipment of treated fuel to a repository, or the emplacement of that treated fuel in the repository—but it does provide some analysis of the impacts of these other components on the selection and implementation of treatment options.

Since the Task Team report was issued in 1996, DOE has undertaken a series of studies to assess the costs and technical viability of implementing one or more of these treatment options. These studies, which are cited in Appendix F of this report, formed the basis for this review. DOE intends to issue a final EIS-record of decision (ROD) in early 1999 that will select one or more of the treatment options for implementation. The findings of the present report will be used as an input to this decision.

The following sections provide a summary of the findings on the three charges in the statement of task given above. Readers interested in a more detailed discussion of these findings should consult Chapters 2-5 of this report.

FINDINGS RELATED TO THE SELECTION
OF ALUMINUM SPENT FUEL TREATMENT OPTIONS

The first charge of the statement of task involves the examination of the set of technologies chosen by DOE for treatment of aluminum spent fuel and the identification of other alternatives that DOE might consider. The first charge of the statement of task is addressed in Chapter 2 through a discussion of four questions, as summarized below:

1. Were a reasonably complete set of treatment alternatives identified by DOE? The answer to this question is a qualified "yes." Although the Task Team apparently made no effort to perform a systematic search for treatment technologies in use in other countries or industries that might be applied to aluminum spent fuel, it appears to have succeeded in identifying a reasonably complete set of alternatives. The affirmative answer to this question is qualified for two reasons. First, the Task Team may have incorrectly eliminated one of the treatment options (chloride volatility treatment) because it believed that no experimental work on this technology had been done. In fact, considerable experimental work on chloride volatility was completed between 1950 and 1965 at Argonne National Laboratory. The P.I. did not attempt to determine whether this treatment option would have ranked above the other advanced treatment alternatives had it been evaluated further by the Task Team. Second, the Task Team did not consider options for treating the depleted uranium spent fuel, which comprises about 40 percent of the inventory of aluminum spent fuel expected to be received at Savannah River. The Task Team identified this fuel as a candidate for reprocessing because the fuel is declad and is comprised of uranium metal that may not be suitable for disposal in the repository. If reprocessing is not possible for policy reasons, however, there is no obvious treatment and disposition pathway for this fuel.

2. *Was the methodology used to screen and rank the treatment alternatives technically sound, and did it lead to the selection of appropriate primary and backup treatment options?* The answer to this question is a qualified "yes." The technical approach used by the Task Team to evaluate and rank the alternative treatment options was appropriate for the degree of technical maturity and the amount and quality of available data, and the ranking methodology was adequate for screening purposes. The output of the ranking methodology appears to be consistent with what one would expect.

The affirmative answer to this question is qualified because neither the Task Team nor other parts of DOE have developed a complete set of process requirements, particularly waste form requirements and other waste acceptance criteria for repository disposal, that would allow a detailed assessment of the treatment options to be made. Most significantly, there appears to be some uncertainty about whether HEU aluminum spent fuel will be acceptable for disposed in a repository because of criticality and nonproliferation concerns. Until such issues are resolved, implementation of appropriate treatment alternatives can not proceed without significant financial and schedule risks. A process that allows for some flexibility and phased decision-making in the selection and implementation of alternative treatment options seems warranted in light of this uncertainty.

3. Are the primary and backup treatment options likely to work as described and produce acceptable waste forms? The direct co-disposal treatment is technically simple and straightforward to implement. The technologies for drying the fuel, placing it into a container, and sealing the container are readily available and can likely be adapted to this application with little additional development work. Melt and dilute treatment is more demanding technically than direct co-disposal treatment and will require a more significant infrastructure, including hot cell space, a melter, and an off-gas treatment system. The radioactive fuel must be melted at temperatures up to about 1000 °C, which will result in the release of some volatile fission products that must be recovered by an offgas system and recycled or otherwise disposed of. All of the technologies needed to make this system function successfully have been used in other applications, and it should be a relatively straightforward exercise to bring them together for aluminum spent fuel treatment. Melt and dilute treatment is worth pursuing despite the additional development and infrastructure requirements because it allows more control over waste form composition and performance characteristics than does direct co-disposal. Additionally, this option would reduce significantly the need for spent fuel characterization and the number of canisters to be interim-stored and eventually shipped to the repository, which would help offset the cost of treatment.

There is not enough information at present about any of the advanced treatment technologies to select a backup option. In particular, not enough is known at present about electrometallurgical treatment, which was selected as the backup option by the Task Team, to determine whether it will work as described, and additional development work will have to be done to determine the feasibility of applying this treatment technology to aluminum spent fuel.

   4. What other treatment options should DOE consider? DOE should have given more careful consideration to the conventional reprocessing option for treating aluminum spent fuel. There appear to be several technical advantages to this option over the others considered by the Task Team.[9] This treatment option has been demonstrated to work for aluminum spent fuel from production reactors, the costs and risks are well known, the necessary facilities are currently in operation at Savannah River, and the waste form (borosilicate glass) will likely be acceptable for disposal at the repository. Reprocessing even a portion of the aluminum spent fuel could significantly reduce the overall costs of treating the total aluminum spent fuel inventory by alleviating the need for additional spent fuel storage space at Savannah River and eliminating the problems with odd-sized fuel elements that may be difficult to process by other methods. There is still some policy uncertainty about whether the F and H Canyons at Savannah River will be available after 2002 for reprocessing, and future developments in DOE proliferation policy may preclude the use of the reprocessing option except in special cases (e.g., disposing of damaged spent fuel elements). It is recommended that DOE-Savannah River undertake a common-basis cost and performance comparison of the two primary treatment alternatives (direct co-disposal and melt and dilute treatment) and conventional reprocessing as part of its process for evaluating and selecting a treatment option.

---

[9] The Task Team was directed to consider alternatives to reprocessing, but the P.I. knows of no restrictions on DOE's ability to develop cost and schedule information on this option, if only for comparative purposes, in the EIS.

FINDINGS RELATED TO WASTE-PACKAGE PERFORMANCE CRITERIA

The second charge in the statement of task involves the examination of the waste-package performance criteria being developed by DOE-Savannah River for aluminum spent fuel and the identification of other criteria that should be considered. Most of the findings relate to the performance criteria that have been developed by DOE-Savannah River in response to the waste acceptance criteria (WAC) published by DOE-Yucca Mountain,[10] but this report also includes comments on other criteria that could have a significant impact on the selection of a treatment option.

The second charge of the statement of task in Chapter 3 was addressed through a discussion of three questions, summarized below:

1. Have all of the important waste-package performance criteria been identified by DOE-Savannah River? The answer to this question is a qualified "yes." DOE-Savannah River staff appear to be working closely with their counterparts at DOE-Yucca Mountain to ensure that the important waste acceptance criteria have been identified and that the right kind of work is being done to demonstrate conformance. DOE-Savannah River also appears to have access to the draft documents being prepared by DOE-Yucca Mountain that could affect the acceptability of aluminum spent fuel for disposal at the repository. The answer is qualified because many of the waste acceptance criteria are preliminary and could change significantly as waste package and repository designs are refined. Continuance of the ongoing dialogue between DOE-Savannah River and

---

[10] The DOE-Office of Civilian Radioactive Waste Management and its management and operating contractor, which are responsible for characterizing the candidate repository at Yucca Mountain.

DOE-Yucca Mountain will be essential to track and respond effectively to any future changes.

   2. *Are there other performance criteria that should be considered?* The answer to this question has three parts, the first for the WAC, the second for the interim storage criteria, and the third for the transportation criteria. The answer to the first part of the second question is "no." The current WAC for the candidate repository at Yucca Mountain are very clearly laid out by DOE-Yucca Mountain documents, and the information received from DOE-Savannah River during the course of this study indicates that all of the potentially applicable WAC have been identified and are being addressed through ongoing work. However, many of the WAC are preliminary and could change significantly as DOE-Yucca Mountain refines the waste package and repository designs. These changes could have significant implications for the acceptability of any waste form for disposal.

   The answer to the second part of the second question—are there other criteria that should be considered for interim storage in addition to those that are required for repository acceptance?—is "no." Most of the criteria seem reasonable given the current plans that DOE-Savannah River has to store, retrieve, and process (as necessary) the fuel to put it into road-ready form. One of the criteria, however, appears to be unnecessary. Specifically, the criterion that sets limits for plastic deformation of the aluminum spent fuel in the disposable canister seems overly restrictive and potentially costly. The justification given for this requirement is that it will provide for ready removal of the fuel from the canister, but it is not clear why DOE-Savannah River would ever want to remove spent fuel from a disposable canister under normal operating conditions, and even under "abnormal" conditions such as a tipover accident the canister could be sectioned to remove the spent fuel. DOE-Savannah River is encouraged to reexamine the cost and potential benefit of this criterion in view of the unlikely need for future fuel removal.

The answer to the third part of the second question—are there other transportation criteria that should be considered?—cannot be answered at this time. It appears that relatively little work has been done to date on establishing criteria to meet transportation requirements. DOE-Savannah River should not encounter any significant problems meeting the requirements in 10 CFR 71, but DOE-Savannah River must review the shipping requirements before it finalizes the design of its disposable canisters.

3. Is the work under way by DOE-Savannah River appropriate to demonstrate conformance with the various criteria and requirements? The answer to this question is a very qualified "yes." The development program under way to demonstrate conformance with the waste acceptance criteria appears to be properly focused and appropriate to the task. This answer is qualified because the short schedule for this project did not allow an in-depth review of all of the ongoing work in the aluminum spent fuel program.

Several of the WAC are poorly defined at present and may be subject to significant future change.[11] It may be quite some time before DOE-Savannah River knows with certainty whether direct co-disposal treatment is viable. This current state of uncertainty has significant implications for the "path forward" for selecting spent fuel treatment options. Three conclusions were identified based on these facts: (1) a single treatment option may not be suitable for all types of aluminum spent fuel; (2) the aluminum spent fuel program will have to maintain flexibility in selecting treatment options until there is more complete information on the WAC and other repository requirements; and (3) a path forward that involves phased decision-making in the selection and implementation of alternative treatment options is indicated.

---

[11] Revised EPA standards, ongoing performance assessment (PA) work at Yucca Mountain, and other developments could result in changes to the WAC.

FINDINGS RELATED TO COSTS AND TIMING OF
ALUMINUM SPENT FUEL TREATMENT OPTIONS

The third and final charge of the statement of task requests that the National Research Council provide—to the extent possible given the accelerated schedule for this project—an assessment of the cost and timing aspects associated with implementation of each aluminum spent fuel treatment option. The four-month schedule for information gathering and report development did not permit a detailed review of the cost and schedule estimates for alternative treatment options. The study focused on a review of the methodologies used to estimate costs to see if they follow generally accepted practices, are applied consistently, and result in estimates that are useful for comparative and programmatic purposes.

The third charge of the statement of task was addressed in Chapter 4 through a discussion of three questions, summarized below:

1. Do the cost estimates account for all of the major cost factors in the aluminum spent fuel treatment program? The answer to this question is "yes." The major cost factors of the system for receiving, treating, handling, storing, and disposing of aluminum spent fuel for each of the treatment options were identified in the Task Team report, and systematic cost estimates for these major cost factors were developed in the alternative cost study report. Both reports provided reasonably complete cost breakdowns, a list of the programmatic assumptions used in the cost estimates, and an explanation of the methodologies used to estimate uncertainties in total system costs.

2. Are the cost and schedule estimates suitable for comparison of the options and selection of one or more preferred alternatives? The answer to this question is a qualified "yes." The cost estimates appear to be sufficiently complete for comparative purposes and for selecting a small number of alternative treatment options for further consideration. However, the schedule estimates for implementing the treatment option range from unrealistic to ambitious, and there is no provision in the cost estimates for additional program delays. Significant program delays will add substantially to the costs for this program.

The answer to this question is qualified because costs did not turn out to be a particularly effective discriminator of the various treatment options, mainly because the treatment options themselves comprised a relatively small part (approximately 20 percent or less) of overall systems costs. There was not much consideration given to reducing overall systems costs by examining alternatives in the fuel receipt and handling schedule. The current schedule appears to be based on current handling and storage capabilities at Savannah River, and relatively little consideration has been given to how changes in this schedule could affect system costs or the selection of alternative treatment technologies.

3. Are the cost and schedule estimates suitable for budget planning purposes? The answer to this question is "no." The schedule estimates are ambitious and depend to a great extent on the timely completion of work by other parts of DOE. The cost and schedule estimates also are limited by the lack of conceptual designs for some of the treatment facilities and because some of the process steps have not yet been demonstrated to work for aluminum spent fuel. Additionally, the cost estimates do not consider the impacts of program delays on costs and schedules. Some amount of delay seems inevitable even under the best of circumstances and could come from several quarters during the budgeting, contracting, construction, and health and safety review phases of the program. DOE-Savannah River has not provided contingencies for such delays in its current program plans.

CONCLUDING OBSERVATIONS

The primary focus of this report is on options for treating aluminum spent fuel. However, spent fuel treatment is just one component of a much larger and complex *aluminum spent fuel disposal program*, a program that is slated to last for about 40 years and cost in excess of $2 billion. The aluminum spent fuel disposal program is a complex web of activities at multiple sites around the world, ranging from operations at foreign and domestic research reactors that generate aluminum spent fuel to the repository development program at the candidate site at Yucca Mountain. Several parties have responsibilities for activities that take place in this program, and the decisions made by one party can have significant impacts on costs, schedules, and current or planned operations elsewhere in the program. DOE-Savannah River must select one or more treatment options for aluminum spent fuel that will meet repository waste acceptance criteria, which have yet to be finalized; design treatment and storage facilities that are sized appropriately to waste streams, which are subject to future change; and provide for interim storage of the processed waste until the repository, which is yet to be designed, licensed, or constructed, is able to accept it.

The spent fuel disposal program is a *systems* problem in the classic sense. It involves several interacting components, each associated with different programmatic factors (e.g., cost, time, safety, policy constraints), multiple responsible parties, and different levels of uncertainty. The selection of aluminum spent fuel treatment options in the face of such uncertainties calls for a phased strategy in which critical programmatic decisions—that is, decisions that involve major program directions and commitments of funds—are made and implemented when the information needed to base sound choices becomes available. The acquisition of information for decision making also is an important part of the phased-strategy approach, both the acquisition of existing data from third-party sources and the generation of new data to fill information gaps. Of course, the phased strategy recognizes that there may be trade-offs between information acquisition and costs of delayed decisions and seeks to maximize the former and minimize the latter.

In the context of the aluminum spent fuel treatment activities at DOE-Savannah River, the primary objectives of the phased strategy should be to maximize the probability of program success, minimize overall costs, and protect the program against the down-side risks from changes over which it has little or no control. The major programmatic decisions that must be made by DOE-Savannah River include the selection of one or more options for treating aluminum spent fuel and also the selection of a design for the treatment, storage, and shipping (TSS) facilities. The criteria for the decision process include the effectiveness of the treatment process, cost, schedule, compliance with applicable environmental health and safety standards, and consistency with other applicable policies. The options selected and facilities constructed also must be matched appropriately to the front (spent fuel generation) and back (disposal and D&D) ends of the overall disposal program.

DOE-Savannah River appears to recognize the importance of a phased decision-making strategy and is already applying it to individual parts of its program. However, a systems-oriented strategy is needed in the treatment program to ensure that technically sound and cost-effective decisions are made and implemented in a timely manner. Three illustrative examples of such strategies are provided below, and a more detailed discussion is given in Chapter 5.

1. Spent Fuel Receipt and Storage. As part of the phased decision strategy on the treatment option for aluminum spent fuel, the fuel receipt and storage schedule will have to be considered, and one of the important programmatic factors on this schedule is the high cost of time.[12] The prompt shipment of all aluminum spent fuel to Savannah River for

---

[12] The cost of time can be thought of as the operational costs that are unrelated to actual production activities. These would include management and administrative costs, costs of supporting workers in a stand-by mode, and other operational costs that are time related rather than production or throughput related, for example, certain types of maintenance costs. To the first order, these operational costs are fixed per unit of time, consequently, cost is approximately proportional to time.

treatment might require the purchase of additional shipping casks but could significantly reduce overall costs. However, shipment and treatment must be phased to minimize the need for new facilities.

    2. Treatment and Interim Storage. Based on information received by the P.I. during the course of this study, there does not appear to be a technical basis for rejecting conventional reprocessing as an option for treating aluminum spent fuel from foreign and domestic research reactors. Conventional reprocessing is a proven and reliable spent fuel treatment technology based on over 300 plant-years of operation worldwide, and the necessary treatment facilities (the F and H Canyons and the Defense Waste Processing Facility [DWPF]) are operating at Savannah River and are being used to treat aluminum spent fuel from research and production reactors.

    The alternative cost study prepared by Westinghouse Savannah River Company suggested that conventional reprocessing was a cost-effective treatment option when compared with direct co-disposal and melt and dilute treatment, the two primary treatment alternatives considered by the Task Team. However, the cost estimates for these three treatment alternatives have not been independently validated in this or any other study. Although it is difficult to make quantitative comparisons between a proven treatment technology such as conventional reprocessing and some of the other unproven treatment technologies considered by the Task Team, it is clear that the cost, performance, and safety of unproven technologies have much greater uncertainties than those of a demonstrated technology such as reprocessing. The common-basis cost and performance comparison of the two primary treatment alternatives (direct co-disposal and melt and dilute treatment) and conventional reprocessing, which was recommended elsewhere in this report, will enable DOE-Savannah River to determine whether conventional reprocessing is an appropriate treatment option for this fuel.

    The concern with conventional processing appears to be mainly one of policy and is related to the use of reprocessing for waste management generally rather than any specific concern about reprocessing this particular fuel type. Current U.S. nonproliferation policy does not encourage the civil use of plutonium. Accordingly, the United States "does not itself engage in plutonium reprocessing for either nuclear power or nuclear explosive purposes."[13] The P.I. notes that plutonium separation is not a significant problem with conventional reprocessing of enriched aluminum spent fuel from research reactors. There is less plutonium in this fuel in comparison to commercial spent fuel owing to its high ^{235}U enrichment, and separation of plutonium is not a required part of reprocessing treatment. The plutonium can be left in the liquid waste stream along with the fission products for later vitrification in glass. For aluminum spent fuel, the ^{235}U separated during conventional processing represents a potential proliferation hazard, but it can be diluted with ^{238}U within the reprocessing facility to make LEU. Moreover, Savannah River is a weapons material secure site and will remain so for the duration of this program.

Finally, the reprocessing of aluminum spent fuel does not appear to be in conflict with the DOE decision to phase out reprocessing at Savannah River. The Highly Enriched Uranium Task Force noted in its predecisional draft report that the need for reprocessing for long-term DOE spent fuel management was unclear at present and that DOE should evaluate the near-term operational requirements to bring its facilities to a condition for transfer to the Office of Environmental Management for potential future operations. Indeed, as noted elsewhere in this report, DOE has or plans to reprocess some of its aluminum spent fuel in the Canyons at Savannah River because of safety concerns.

It is the P.I.'s opinion that the acceptability of conventional reprocessing might be increased if it were redesigned as a reprocess-and-

---

[13] The quote is taken from the White House Fact Sheet entitled "Nonproliferation and Export Control Policy" dated September 27, 1993. The fact sheet is based on Presidential Decision Directive 13, which is classified and was not reviewed in this study.

dilute operation, in which spent fuel is conventionally reprocessed and the separated ^{235}U is diluted with ^{238}U to produce low-enriched uranium before it leaves the reprocessing facility.[14]

   3. Post-2015 Aluminum Spent Fuel Inventory. There does not appear to be any reason at this time to make a decision about the disposition of the post-2015 inventory of aluminum spent fuel. This inventory could well be different in size (most likely smaller) and composition than currently anticipated, therefore, treatment options that do not appear to be available today may in fact be available when the time comes to treat this fuel.

   DOE-Savannah River is doing a commendable job of collecting data for decision making on many of the individual components of its treatment option selection program. In addition, DOE-Savannah River is in the process of defining a decision strategy for selecting and implementing a treatment option for aluminum spent fuel. As part of this decision strategy, it is recommended that DOE-Savannah River conduct a complete systems review to identify and understand the relationships among the various components of the aluminum spent fuel disposal program. DOE-Savannah River also is encouraged to apply a phased strategy for selecting and implementing a treatment option for aluminum spent fuel that takes into account the considerations discussed above. This phased approach will support the analysis required in the environmental impact statement and will lead to a more credible EIS-ROD and a more successful and cost-effective path forward for the aluminum spent fuel treatment program.

---

[14] This dilution could in fact be done at almost any step of the process.

1
BACKGROUND AND TASK

The Office of Environmental Management (EM) within the U.S. Department of Energy (DOE) is responsible for managing and preparing for disposal all noncommercial spent nuclear fuel generated by U.S. nuclear materials and research reactor programs as well as some foreign research reactor programs. EM has responsibility for more than 150 different fuel types in storage at DOE and research reactor facilities around the world. This fuel exists in a number of different chemical forms (e.g., metal oxide, aluminum alloy, carbide) with different types of metal cladding (e.g., zircaloy, aluminum, stainless steel) and with uranium-235 (^{235}U) enrichment levels ranging from a few percent to more than 90 percent (DOE, 1995). At the end of 1995, DOE EM had under management approximately 2,650 MTHM[1] of spent nuclear fuel, primarily at the Hanford, Idaho National Engineering and Environmental Laboratory (INEEL), and Savannah River sites (DOE, 1996a; Figure 1.1).[2]

Beginning in 1995, DOE completed a number of programmatic environmental impact statements (PEISs) and issued a series of records of decision (RODs) affecting the management of spent fuel throughout the DOE complex. The Savannah River Site, which is located near Aiken, South Carolina (Figure 1.1), was designated to manage the department's

---

[1] Metric tons heavy metal, the amount of heavy metal (uranium, thorium, and plutonium) present in fresh (unirradiated) fuel.

[2] Although the quantity of DOE spent nuclear fuel is small compared to commercial inventories, which totaled about 34,000 MTHM at the end of 1995 (DOE, 1996a), the DOE fuel is more variable in chemical form, ^{235}U enrichment, and physical size.

inventory of aluminum spent nuclear fuel[3] from foreign[4] and domestic[5] research reactors, the subject of this report.[6] During the next four decades, Savannah River will be responsible for receiving and storing this fuel at the site, processing it as necessary to put it into "road-ready" form[7] for eventual shipment to a repository, and providing for interim storage of the road-ready product until a repository is ready to accept it.

Approximately 20 MTHM of the aluminum spent fuel from research reactors was in storage at Savannah River as of October 1, 1997 (WSRC, 1998): 19 MTHM of depleted uranium[8] targets and blankets and 1 MTHM of highly enriched uranium (HEU).[9] This fuel is being stored in

---

[3] Aluminum spent nuclear fuel is irradiated fuel that contains uranium-aluminum matrix fuel elements and (or) is clad in aluminum. DOE refers to this fuel as "aluminum-based" or "aluminum-clad" spent fuel, but in this report this fuel will be referred to as "aluminum spent fuel."

[4] Under the Atoms for Peace Program begun in the 1950s, the U.S. government supplied research reactor technology and nuclear fuel to foreign nations that agreed to forgo the development of nuclear weapons. In 1996, DOE completed an EIS on the management of this foreign reactor fuel (DOE, 1996c) and in a subsequent ROD decided to accept and manage foreign research reactor spent fuel in the United States.

[5] The domestic research reactors are operated by DOE, other federal agencies, and universities.

[6] Savannah River also is responsible for the management of aluminum spent nuclear fuel from production reactors, which were used for nuclear weapons materials production. This spent fuel is being treated by conventional reprocessing (i.e., wet chemical processing) in the F and H Canyons at Savannah River and is not considered in this study.

[7] That is, the fuel will be treated and loaded into disposable canisters suitable for interim storage, shipping, and loading at the repository site into a final repository package.

[8] Depleted uranium is depleted in ^{235}U relative to natural abundances.

[9] Defined as a material that contains more than 20 percent ^{235}U by mass. In contrast, natural uranium contains about 0.7 percent ^{235}U by mass.

FIGURE 1.1 Location of DOE sites with large inventories of spent nuclear fuel. SOURCE: Modified from DOE (1996b), Figure 1-2.

two water-filled pools, or wet basins, at the site: the L-Reactor Basin,[10] (L-Basin for short) and the Receiving Basin for Offsite Fuels, (RBOF; see Figure 1.2).

By 2035, the site expects to receive an additional 10 MTHM from government and university reactors around the United States and 12

---

[10] The L-Reactor is one of several production reactors at Savannah River used to produce nuclear weapons materials. This reactor has been shut down but the storage pool (the L-Basin) is still in operation.

FIGURE 1.2 Interior view of the RBOF facility. After receipt at the facility, the fuel is placed into racks in the water-filled pool or "wet basin" shown in the photo. The water surface can be seen on the right side of the pool. The water is conditioned to minimize aqueous corrosion of the fuel. SOURCE: Savannah River Laboratory.

MTHM from research reactors in more than 25 countries (Figures 1.3, 1.4).[11] The domestic fuel is expected to include 9.4 MTHM of HEU and 0.6 MTHM of LEU,[12] whereas the foreign fuel is expected to include 2.6 MTHM of HEU and 9.4 MTHM of LEU. Some of the fuel that is planned to be received has not yet been fabricated or shipped to the reactors where it will be used. The inventory of materials to be received is very diverse, ranging from intact reactor fuel elements to damaged or corroded fuel elements to uranium-bearing materials such as reactor blankets and powdered irradiation targets (Figure 1.5).

DOE is now in the process of developing a draft PEIS for management of aluminum spent fuel. This PEIS will determine the need for additional treatment and storage facilities at the site to accommodate the receipt of fuel from domestic and foreign sources. It also will assess alternative treatment technologies and select one or more treatment options for preparing the fuel for disposal in a repository. Formal decisions on treatment options and facilities are expected to be announced in a ROD that will be issued by DOE in the first quarter of FY99. Implementation[13] of the ROD is expect to begin around 2000.

TASK STATEMENT AND STUDY PROCESS

As part of its efforts to prepare the aluminum spent fuel PEIS, the Savannah River Office of DOE asked the National Research Council (NRC) to review the options that it has identified to treat aluminum spent fuel (see Appendix A). These options are discussed in Chapter 2 of this report.

---

[11] The future amount of aluminum spent fuel to be received by Savannah River is only an estimate and was provided by DOE-Savannah River staff.

[12] Low-enriched uranium, which is defined as material that contains 20 percent or less ^{235}U by mass.

[13] Implementation is defined by DOE as "an authorized project underway (i.e., detailed design begun)" (WSRC, 1998).

FIGURE 1.3. (A) Locations of domestic research reactors that have or will generate aluminum spent fuel for shipment to and treatment at Savannah River. SOURCE: Data from Savannah River Laboratory.

The statement of task for this study involves a review of the following aspects of DOE's program for developing a preferred strategy to treat for disposal this aluminum spent fuel:

 • examination of the set of technologies chosen by DOE and identification of other alternatives that DOE might consider;
 • examination of the waste-package performance criteria developed by DOE to meet anticipated waste acceptance criteria for disposal of aluminum spent fuel and identification of other factors that DOE might consider; and

FIGURE 1.3. (B) Locations of foreign research reactors that have or will generate aluminum spent fuel for shipment to and treatment at Savannah River. SOURCE: Data from Savannah River Laboratory.

•   to the extent possible given the schedule for this project, an assessment of the cost and timing aspects associated with implementation of each spent nuclear fuel treatment technology.

This study is focused primarily on the treatment step of the process for disposing of aluminum spent fuel, that is, the options for treating this fuel to make it acceptable for disposal in a repository. This report does not review the other components of the disposal program—for example, the shipment of aluminum spent fuel to Savannah River for treatment, the shipment of treated fuel to a repository, or emplacement of

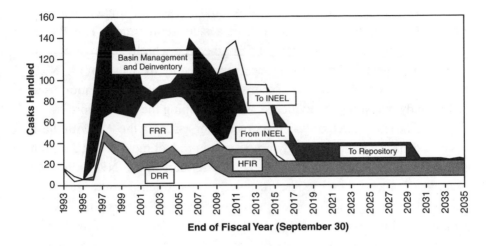

FIGURE 1.4 Projected receipts and transfers of aluminum spent fuel at Savannah River. NOTE: FRR = foreign research reactor fuel; DRR = domestic research reactor fuel; HFIR = High-Flux Isotope Reactor fuel (HFIR is located at the Oak Ridge site; see Figure 1.1); from INEEL = aluminum spent fuel now in storage at the Idaho National Engineering and Environmental Laboratory that will be shipped to Savannah River for treatment; to INEEL = non-aluminum spent fuel at Savannah River that will be shipped to INEEL for treatment; to repository = treated aluminum spent fuel that will be shipped to a repository for disposal; SRS = Savannah River Site. SOURCE: WSRC (1998), Figure 2.

that treated fuel in the repository[14]—but it does provide some analysis of the impacts of these other components on the selection and implementation of treatment options.

At the request of DOE, this study was undertaken using a principal investigator (P.I.) rather than a study committee. The P.I., a member of the National Academy of Engineering who has a great deal of expertise in spent fuel processing, was appointed by the Chair of the NRC to perform this study. He was responsible for gathering information used in the study, weighing the evidence, and developing the final report.

The study director was responsible for assisting the P.I. with these tasks and ensuring that the study process and product met the NRC's high standards of objectivity and completeness. As part of the NRC's quality assurance process, the report received a thorough review prior to publication by a group of experts selected by the NRC's Report Review Committee for their knowledge of the issues under consideration. The review was conducted anonymously, that is, the identities of the reviewers were not divulged to the P.I. until after the report was approved for publication. The report reviewers are identified in the Foreword.

Information for this study was gathered in two meetings that were open to the public: the first held in Aiken, South Carolina, on November 4-5, 1997, and the second held in Augusta, Georgia on December 2-3, 1997. The agendas for these meetings are given in Appendix B. During the first meeting, the P.I. and study director received scoping briefings from DOE and contractor staff on the aluminum spent fuel program at Savannah River and the work completed to date on the alternative treatment technologies.

The second meeting comprised a more detailed set of briefings on the alternative treatment technologies and was attended by the P.I.,

---

[14] For the purposes of this study it was assumed that the treated aluminum spent fuel waste form would be subject to standard repository procedures and that the fuel waste form would be packaged in the outer container being developed for other repository waste.

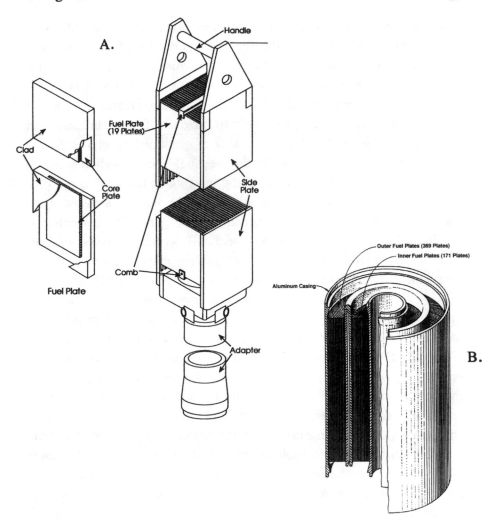

FIGURE 1.5 Examples of some of the spent research reactor fuel to be treated at Savannah River: (A) Materials and Test Reactor fuel type, which is used in many research reactors. (B) High-Flux Isotope Reactor fuel from Oak Ridge. SOURCE: Task Team (1996).

study director, and 11 expert consultants (hereafter referred to as "consultants") selected by the P.I. for their knowledge of and experience with the following issues: metallurgy and corrosion, remote processing, criticality, proliferation, cost, and regulations. Each consultant was charged with reviewing the issues within his or her own area of expertise but was not asked to evaluate the overall program. The consultants were asked to provide written answers to sets of questions developed by the P.I. and the study director prior to the meeting. These question sets also were provided to DOE staff prior to the meeting so that they could plan informative presentations. The questions and the written answers of the consultants are provided in Appendixes C and D, respectively. Additionally, the P.I. solicited and received written answers to these questions from two other consultants on proliferation policy who did not attend the second meeting. These also are included in Appendix D.

The findings, conclusions, and recommendations in the body of this report are the responsibility of the P.I., who used his professional judgment in weighing the evidence presented by DOE and the consultants (Appendix D), as well as the responsibility of the NRC, which, as noted previously, subjected the report to review by an independent group of reviewers. The P.I. considered carefully the advice he received from the consultants in developing this report, and in many cases the findings, conclusions, and recommendations in the report body are consistent with those articulated by the consultants in their individual reports (Appendix D). In some cases, however, the P.I. reached different conclusions than the consultants on particular issues—because, for example, there was no agreement among the consultants on a particular issue or because overall system considerations lead the P.I. to a different conclusion. The consultants' reports have been reproduced in full in Appendix D so that readers can assess for themselves the advice offered therein, and also so that the process used in this review will be transparent and differences among consultants and the P.I. will be readily apparent.

ORGANIZATION OF THIS REPORT

The remainder of the report is organized into four chapters, one chapter for each of the three charges in the statement of task and a concluding chapter with additional observations judged to be of value to DOE's program. An effort was made to keep the chapters as short as possible but also to provide enough information so that they should be understandable to nonexperts. Each chapter begins with background material and data needed to understand the issue and concludes with discussions of each charge of the task statement.

2

TREATMENT OPTIONS FOR ALUMINUM
SPENT NUCLEAR FUEL

This chapter focuses on the first charge in the statement of task, "an examination of the set of technologies chosen by the Department of Energy (DOE) and identification of other alternatives that DOE might consider" for treating aluminum spent fuel.[1] To address this charge, this chapter provides a review of the options for treating aluminum spent fuel that were identified by DOE, as well as the methodologies used by DOE to rank these options and identify primary and backup technologies.

The main sources of information used in this assessment are the presentations received from DOE and contractor staff at the two information-gathering meetings (see Chapter 1 and Appendix B) and the following reports provided by DOE and Westinghouse Savannah River staff:

- *Technical Strategy for the Treatment, Packaging, and Disposal of Aluminum-Based Spent Nuclear Fuel* (Task Team, 1996), and
- *Alternative Aluminum Spent Nuclear Fuel Treatment Technology Development Status Report* (WSRC, 1997a).

The first part of this chapter provides a brief review of treatment alternatives and the methodologies used by DOE to select the primary and backup treatment options. The second part of the chapter provides an analysis of this work in response to the first charge of the statement of task.

---

[1] As noted in Chapter 1, the term "aluminum spent fuel" refers to aluminum-clad or aluminum-matrix spent nuclear fuel from foreign and domestic research reactors, the subject of this report.

BACKGROUND

In 1995, the Office of Spent Fuel Management of DOE established a team of experts (the Research Reactor Spent Nuclear Fuel Task Team, hereafter referred to simply as the "Task Team") to help develop a technical strategy for treatment of aluminum spent fuel to put it into "road ready" form for disposal at a repository. The Task Team comprised ten core staff and seven technical support staff drawn from the DOE complex and associated organizations. These core and support staff had expertise in spent fuel behavior and treatment, fuel handling, nuclear criticality, regulation and licensing, waste-package design, and repository requirements.

The Task Team was asked to evaluate alternative treatment and packaging technologies that could be used in the place of conventional reprocessing[2] to interim store and ultimately dispose of aluminum spent fuel in a safe and cost-effective manner (Task Team, 1996, p. 2). The team produced a report that evaluated various treatment technologies and recommended a strategy for selecting, developing, and implementing[3] one or more treatment options by the year 2000. The Task Team report was an important input to the current study, and the chair of the Task Team, Mr. Jack DeVine of Polestar Applied Technology, Inc., provided a presentation of the Task Team's work at the second information-gathering session for this study.

As noted above, the focus of the Task Team's work was on technologies for treating aluminum spent fuel *that could be used in the place of conventional reprocessing.* Savannah River currently has two reprocessing facilities in operation (the F and H Canyons; see Figure 2.1)

---

[2] Conventional reprocessing involves the dissolution of aluminum spent fuel in acid followed by the chemical recovery of uranium. The remaining liquid waste stream, which contains plutonium (if present in the spent fuel) and fission products, is stabilized by forming it into glass in a process known as vitrification.

[3] As noted in Chapter 1, DOE now defines implementation as "an authorized project underway (i.e., detailed design begun)" (WSRC, 1998).

FIGURE 2.1 Oblique aerial view of the F Canyon chemical separations plant at Savannah River where spent fuel is separated into its constituent components, including uranium, fission products, and, if desired, plutonium. SOURCE: Savannah River Laboratory.

and a facility for making high-level waste glass (the Defense Waste Processing Facility, or DWPF) that are being used to reprocess fuel and targets from the production reactors at Savannah River. In principle, the Canyons also could be used to process the inventory of aluminum spent fuel from research reactors, which is small in comparison to the inventory of aluminum spent fuel from production reactors. In fact, some aluminum spent fuel from research reactors has been (or will be) processed in the Canyons because it was either degraded, declad, or perceived to represent a higher risk than other parts of the inventory. The Task Team identified approximately 35 MTHM of fuel[4] that potentially could be reprocessed for these reasons (Task Team, 1996, Table 5.2-1). This material includes damaged, failed, and sectioned fuels, particulate target materials, and depleted uranium blanket fuels.

The need to develop alternative treatment technologies for aluminum spent fuel was necessitated by DOE's policy to phase out reprocessing. The policy to phase-out reprocessing at DOE sites was announced by the Secretary of Energy in February 1992 (DOE, 1992a), based on recommendations from the Highly Enriched Uranium Task Force.[5] The phaseout of the reprocessing Canyons at Savannah River is now scheduled to occur at about 2002. Prior to phaseout, those materials identified as "at risk," including some aluminum spent fuel from research reactors, will be reprocessed in the Canyons (Figure 2.2).

In addition, DOE is currently undertaking a review of the proliferation risks associated with the disposal of foreign research reactor fuel[6] and is expected to issue a report in the second quarter of fiscal year

---

[4] Some of this fuel already has been processed, including the Taiwan Research Reactor fuel that was processed in H Canyon because of the poor condition of its cladding.

[5] A predecisional draft of the Highly Enriched Uranium Task Force report on DOE spent fuel reprocessing was completed in February 1992 (DOE, 1992b). A copy of this draft was not made available to the P.I. until the present report was in the final stages of review.

[6] The foreign research reactor spent fuel EIS (DOE, 1996c) called for DOE to commission or conduct an independent study of the nonproliferation and other implications of reprocessing spent nuclear fuel from foreign research reactors. A study is being conducted by DOE's Office of Arms Control and Nonproliferation, and a staff member from that office (Jon Wolfstol) provided a briefing of this activity at the second information-gathering meeting.

FIGURE 2.2 At-risk materials include spent fuel with damaged or degraded cladding as shown here, for example, by the corrosion pits on the cladding of a Materials and Test Reactor fuel element. The pits are about 1 cm in diameter. SOURCE: Savannah River Laboratory.

1998 (WSRC, 1998; see Chapter 5). According to Savannah River staff, this study also may address reprocessing of domestic research reactor fuel. Because the receipt of aluminum spent fuel currently is scheduled to continue for several decades beyond the scheduled operation of the Canyons at Savannah River, non-reprocessing treatment options for this spent fuel will have to be planned unless the present schedules are revised.

TREATMENT TECHNOLOGIES FOR
ALUMINUM SPENT FUEL

The Task Team evaluated 11 treatment technologies for aluminum spent fuel identified in the foreign research reactor spent fuel environmental impact statement (EIS; DOE, 1996c) and the DOE spent fuel technology integration plan (DOE, 1996b).[7] The treatment technologies evaluated by the Task Team fell into one of three categories:

1. *Direct Disposal Technologies.* These involve no processing of the spent fuel except for drying.[8] After the fuel is dried, it would be placed in a disposable canister and stored until it could be shipped to a repository.

2. *Highly Enriched Uranium* (HEU) *Dilution Technologies.* In these, the fuel is physically or chemically diluted with depleted uranium to reduce the concentration of uranium-235 (^{235}U) to 20 percent or less by mass. Both criticality and proliferation risks are perceived by many to be lower for low-enriched uranium (LEU) than for HEU.

3. *Advanced Treatment Technologies.* Here, the fuel is processed to produce other waste forms for repository disposal than is possible for either the direct disposal or the HEU dilution technologies and to reduce the volume of waste for disposal.

A brief review of the technologies considered by the Task Team is given below. Readers interested in obtaining more detailed descriptions should consult the Task Team report (Task Team, 1996).

---

[7] At the second information-gathering meeting for this study Mr. Jack DeVine, the chair of Task Team, commented that a solicitation for treatment ideas also was distributed to DOE laboratories.

[8] At present, all of the aluminum spent research reactor fuel at Savannah River is being stored underwater.

Direct Disposal Technologies

Two treatment options were considered by the Task Team, *direct disposal treatment*[9] and *direct co-disposal treatment*. A third treatment option, *can-in-canister treatment*, was eliminated during an initial screening step.[10]

Direct Disposal Treatment. The fuel assemblies are trimmed to remove nonfuel components (e.g., end fittings and grappling fixtures), dried, and loaded into canisters to meet the ^{235}U mass limits imposed by transportation or repository criticality requirements (each canister could contain several fuel assemblies). Neutron poison inserts (e.g., borated steel) are added to the canisters as needed to reduce the criticality hazard or to allow increased fuel loading. According to the Task Team, approximately 1,100 canisters would be required to dispose of all of the aluminum spent fuel to be sent to the repository.

Direct Co-Disposal Treatment. The fuel assemblies are cropped as required, dried, and loaded into canisters that will fit into the center spaces of repository waste packages containing high-level waste glass logs produced in the DWPF. A conceptual design for the repository waste package is shown in Figure 2.3. The Task Team estimated that approximately 1,400 canisters would be required to dispose of all of the aluminum spent fuel to be sent to the repository.

---

[9] The Task Team's use of the term *direct disposal* to refer to both the class of treatment technologies and a specific treatment option is potentially confusing. We use the suffix *technologies* (i.e., direct disposal technologies) when referring to the class of treatment options and the suffix *treatment* (i.e., direct disposal treatment) when referring to the specific treatment option to alleviate confusion.

[10] In this option, spent fuel would be loaded into cans, which in turn would be placed in stainless steel canisters and encapsulated in high-level waste glass. The Task Team eliminated this option because of the technical difficulties associated with the low melting point of aluminum spent fuel.

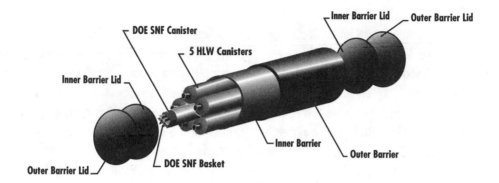

FIGURE 2.3 Schematic illustration of a co-disposal package. The disposable canister that contains the aluminum spent fuel is packaged in a double barrier, corrosion-resistant container with five high-level waste glass logs. NOTE: HLW = high level waste; SNF = spent nuclear fuel. SOURCE: Task Team (1996) Figure 1.3-1.

HEU Dilution Technologies

The Task Team also considered two dilution technologies: *press and dilute treatment* and *melt and dilute treatment*. A third option, *chop and dilute treatment*, was eliminated during an initial screening step.[11]

---

[11] In this treatment option, the spent fuel is shredded mechanically and combined with depleted uranium. The Task Team eliminated this treatment option from

These treatment technologies have three major advantages over the direct disposal technologies: (1) the volume of waste and, hence, the number of waste packages are reduced, and (2) the perceived criticality potential may be lowered; and (3) the proliferation or security risk may be lowered.

Press and Dilute Treatment. The dried and sized aluminum spent fuel assemblies are physically pressed into sandwiches along with sheets of depleted uranium to produce dimensionally uniform packages with composite ^{235}U enrichments of 20 percent or less by mass.[12] Neutron poisons would be added if necessary to further reduce the criticality potential of the fuel. The sandwiches would then be loaded into canisters that will fit into the center space of a repository waste package containing high-level waste glass logs produced in the DWPF (e.g., Figure 2.3) in the same manner as proposed for direct co-disposal treatment. At a 20 percent composite ^{235}U enrichment, about 400 canisters would be required to dispose of all of the aluminum spent fuel to be sent to the repository.

Melt and Dilute Treatment. The aluminum spent fuel is combined with depleted uranium and melted in a crucible to produce an alloy that has a ^{235}U enrichment of 20 percent or less by mass. The material is solidified in a mold (the mold may be the crucible itself) and is loaded into canisters for co-disposal with the glass logs produced in the DWPF (e.g., Figure 2.3). At a 20 percent composite ^{235}U enrichment, about 400 canisters would be required to dispose of all of the aluminum spent fuel to be sent to the repository.

---

consideration because the other dilution treatment options were similar in design and were deemed to be superior.

[12] As noted previously, a ^{235}U enrichment level of 20 percent or less by mass is considered "low enriched" and is less susceptible to criticality and proliferation.

Advanced Treatment Technologies

Five advanced treatment options were considered by the Task Team: (1) plasma arc treatment, glass material oxidation and dissolution treatment, dissolve and vitrify treatment, electrometallurgical treatment, and processing and co-disposal treatment. A sixth treatment option, chloride volatility treatment, was eliminated from further consideration during an initial screening step.[13]

Plasma Arc Treatment. For this treatment concept, spent fuel is combined with depleted uranium and melted in a plasma arc furnace at high temperature to produce a vitreous ceramic with a uranium enrichment of 20 percent or less by mass. The Task Team estimated that at a 20 percent composite ^{235}U enrichment, about 400 canisters would be required to dispose of all of the aluminum spent fuel to be sent to the repository. These canisters could be co-disposed with the glass logs produced in the DWPF. This option is similar to melt and dilute treatment except that the melting takes places at a higher temperature, which allows the fuel to be oxidized to produce a ceramic waste form. The Task Team assumed that a borosilicate glass waste form also could be produced from the plasma arc waste products through a multi-step process, with the later steps occurring at more modest temperatures.

Glass Material Oxidation and Dissolution Treatment. The fuel is placed into a melter with depleted uranium, lead oxide is added to oxidize the metals, and a frit (essentially a powdered borosilicate glass) is added to make glass with a ^{235}U enrichment of 20 percent or less by mass. The lead is recovered from the melt and reused. The Task Team estimated that

---

[13] In this option, the spent fuel would be reacted with chlorine gas at elevated temperatures to produce volatile chlorides, which would be separated and recovered by scrubbing and fractional distillation. The Task Team eliminated this option from consideration because it claimed that no experimental work had been done. As noted later in this chapter, however, that claim is incorrect.

about 800 glass logs would be produced to dispose of all of the aluminum spent fuel to be sent to the repository. These logs would be co-disposed with the glass logs produced in the DWPF.

Dissolve and Vitrify Treatment. The fuel is dissolved in acid with enough depleted uranium to reduce the ^{235}U concentration to 20 percent or less by mass. The solution is transferred to a vitrification plant where it is combined with frit to produce glass logs. The Task Team estimated that about 800 DWPF-size glass logs[14] would be produced to dispose of all of the aluminum spent fuel to be sent to the repository. The logs would be co-disposed with the glass logs produced in the DWPF.

Electrometallurgical Treatment. The aluminum spent fuel is melted to produce a metal ingot. The ingot is then placed in an electrorefiner, and the aluminum, uranium, and fission products are separated. The fission products are oxidized and dissolved in glass to produce about 90 DWPF-size glass logs. The recovered uranium is remelted and combined with depleted uranium to produce an ingot with a ^{235}U enrichment of 20 percent or less by mass, which could be used to make commercial fuel.[15]

Processing and Co-Disposal Treatment.[16] The Task Team referred to this option as a "reference technology" because the reprocessing component of this option has well-known cost and

---

[14] Cylindrical logs having a diameter of 2 feet and a length of 10 feet.

[15] Note that the recovered uranium will contain uranium-236 (^{236}U), which is produced by ^{235}U neutron capture during irradiation. ^{236}U is a neutron absorber, and if its concentration in the recovered uranium is significant, that uranium will have to be downblended less to obtain a higher concentration of ^{235}U to be usable in a reactor.

[16] "Processing" in this context is conventionally referred to as "reprocessing" as defined in footnote 2 of this chapter. We defer to the Task Team terminology when describing this treatment option but use the term "conventional reprocessing" to describe this process elsewhere in the report.

performance characteristics and therefore could be used as a baseline against which the other options could be compared. The Task Team assumed that a portion of the aluminum spent fuel would be reprocessed in the Savannah River Canyons to recover the uranium, which subsequently would be diluted with depleted uranium to obtain a ^{235}U enrichment of 20 percent or less by mass and sold for commercial purposes. The remaining liquid waste, which contains small amounts of plutonium and the fission products, would be fed into the DWPF to produce about 120 glass logs, which would be co-disposed in the repository. The portion of the spent fuel not treated by reprocessing (i.e., the portion of fuel received at Savannah River after the Canyons are shut down) would be subject to other treatments.

EVALUATING THE TREATMENT ALTERNATIVES

The Task Team used a combination of qualitative and quantitative methodologies to screen and rank the alternative treatment options. The objective of this exercise was to eliminate from further consideration those options that were less likely to be implemented successfully because of technical, cost, or scheduling difficulties. The Task Team used four ranking criteria based on the following requirements:

1. The foreign research reactor spent fuel EIS (DOE, 1996c) establishes a target date of 2000 for implementation of an alternative technology for aluminum spent fuel. Thus, the treatment option must have a fair chance of being implemented by the year 2000.[17]

2. The treatment technology must produce a waste form that is acceptable for disposal in a repository. Thus, the waste form must

---

[17] This target date is unrealistic for all but conventional reprocessing if "implemented" means that treatment must be under way. As noted in footnote 3 of this chapter, however, the term has been defined by DOE to mean that a detailed design of the treatment process is under way.

conform to waste acceptance criteria being developed for Yucca Mountain (see Chapter 3).

    3. The treatment technology must meet existing health, safety, and environmental requirements.

    4. The cost of developing the alternative technology must fit within expected DOE budgets. The Task Team determined that no more than about $500 million would be available for development over a five-year period.

    The four ranking criteria used by the Task Team are (1) confidence in success, (2) cost, (3) technical suitability, and (4) timeliness. A brief description of these criteria is given in Table 2.1. The Task Team assigned a weighting factor to each of these screening criteria as shown in Table 2.1. A composite score for each treatment alternative was calculated by multiplying each weighting factor by a scaling factor between 1 and 10, which was assigned by expert judgment, and then summing the results. The results of the rankings are shown in Table 2.2. The overall scores ranged from 74 for direct co-disposal treatment to 23 for plasma arc treatment. The three most highly ranked alternatives were direct co-disposal treatment, melt and dilute treatment, and press and dilute treatment. No score was calculated by the Task Team for processing and co-disposal treatment, because, as noted previously, the team was directed to consider alternatives to conventional reprocessing.

    The Task Team performed a simple sensitivity analysis to determine whether the relative rankings would change if any of the ranking criteria were eliminated from the analysis. To this end, the Task Team recalculated the rankings after removing from the analysis, one at a time, each of the four evaluation criteria shown in Table 2.1. The results of this analysis are shown in Table 2.3. The Task Team observed that the rankings did not change significantly by removing any one of the criteria, suggesting that the overall results were not dominated by any single assumption, criterion, or evaluation.

Table 2.1 Screening Criteria and Weighting Factors Used by the Task Team to Evaluate Treatment Options

Evaluation Category	Screening Criteria	Weighting Factor	Scaling Factor
Confidence in success—evaluators' confidence that the technology can meet technical performance requirements at predictable cost and schedule	Technology development must be reasonably complete by 2000	30	1-10 scale reflecting evaluators' judgment of the likelihood of success for each candidate treatment option, a higher likelihood of success yields a higher score
Cost—projected total implementation costs	Costs must be less than $500 million for the first 5 years	30	1-10 scale; higher-cost options receive lower scores
Technical suitability—technical merits of each technology	Final waste form must meet anticipated repository waste acceptance criteria, and technology must meet environmental, health, and safety requirements	20	1-10 scale reflecting evaluators' judgment, a higher suitability yields a higher score
Timeliness—projected time to implement technology		20	1-10 scale, with the highest score (10) assigned to any technology projected to be in place and operating by 2001

SOURCE: Task Team (1996), Table 4.4-1.

TABLE 2.2 Results of the Task Team Evaluation of Treatment Options for Aluminum Spent Fuel

Treatment Option	Scaling Factor for Each Criterion in Table 2.1[a]				Overall Score[b]
	Confidence in Success (30%)	Cost (30%)	Technical Suitability (20%)	Timeliness (20%)	
Direct co-disposal	6	10	5	8	74
Melt and dilute	6	8	7	6	66
Press and dilute (20%)[c]	6	7	7	6	63
Press and dilute (2%)[d]	7	5	6	6	62
Direct disposal	5	7	4	8	60
Electrometallurgical	3	5	8	3	46
Dissolve and vitrify	4	1	7	1	31
GMODS[e]	1	2	7	1	25
Plasma arc	1	2	6	1	23

[a] The scaling factors range from 1 to 10 for each criterion, where 1 is least suitable and 10 is most suitable.

[b] The overall score was calculated by multiplying each of the scaling factors by the weights shown in Table 2.1, summing the results, and multiplying by 100. The range of possible scores is 10 to 100, where 100 is the most suitable treatment option of those considered. Although the score is quantitative, it is based on qualitative evaluations by the Task Team.

[c] Dilution to 20% ^{235}U by mass.

[d] Dilution to 2% ^{235}U by mass.

[e] Glass material oxidation and dissolution.

SOURCE: Task Team (1996), Table 4.4-2.

TASK TEAM RECOMMENDATIONS

The Task Team noted that although all of the technologies it considered were potentially capable of converting aluminum spent fuel into an acceptable waste form for disposal, no single technology appeared to be optimal for all fuel types or sufficiently mature to be relied on to the exclusion of others.[18] Therefore, the Task Team recommended that DOE continue to examine and develop several treatment options to maintain flexibility and increase the overall likelihood of success.

The Task Team recommended direct co-disposal treatment as the primary treatment option, with melt and dilute treatment as a "parallel" option. The task team noted that direct co-disposal treatment "is the simplest of the technology options evaluated, and seems technically achievable in all respects, at moderate cost and on a timetable consistent with DOE's needs" (Task Team, 1996, p. 68). However, the Task Team recognized that the acceptability of HEU in the repository was a major uncertainty for the success of direct co-disposal treatment, and noted that the waste form produced by melt and dilute treatment was potentially more acceptable and probably more easily licensed.

The Task Team also recommended electrometallurgical treatment, which was the highest-scoring advanced treatment technology (Table 2.2), as a backup option, because the technology is fundamentally different from the others and thus offers some protection against unforeseen technical or licensing problems. The Task Team also noted that the borosilicate glass waste form, the product of this treatment option, is "very robust and highly likely to meet the regulatory requirements . . ." (Task Team, 1996, p. 68).

Finally, the Task Team recommended that DOE focus its development funds on direct co-disposal treatment and melt and dilute

---

[18] As noted previously, the Task Team did not consider conventional reprocessing in its evaluation.

TABLE 2.3 Results of Sensitivity Study to Determine Significance of
Various Evaluation Criteria in Table 2.1

Technology	Sensitivity Case Rankings[a]				
	Base	Case 1	Case 2	Case 3	Case 4
Direct disposal	5	3	5	3	5
Direct co-disposal	1	1	2	1	1
Press and dilute (20%)	3	3	3	4	3
Press and dilute (2%)	4	5	1	5	4
Melt and dilute	2	2	3	2	2
Plasma arc	9	8	9	8	9
GMODS[b]	8	7	8	8	8
Dissolve and vitrify	7	9	7	7	7
Electrometallurgical	6	6	6	6	6

[a] Rankings range from 1 to 9, with a lower number indicating a higher
ranking.

[b] Glass material oxidation and dissolution.

NOTE: Base case: Rankings with all screening criteria (Table 2.1) included;
Case 1 = confidence in success excluded; Case 2 = cost excluded; Case 3 =
technical suitability excluded; Case 4 = timeliness excluded.

SOURCE: DOE Task Team (1996) Table 4.4-3.

treatment, and that it follow the work being done elsewhere (at Argonne
National Laboratory) on electrometallurgical treatment.

RESPONSE TO FIRST CHARGE IN STATEMENT OF TASK

The first charge of the statement of task involves an examination
of the set of technologies chosen by DOE for treatment of aluminum
spent fuel and to the identification of other alternatives that DOE might

consider. As noted previously in this chapter, the Task Team selected direct co-disposal treatment and melt and dilute treatment as the primary treatment options and electrometallurgical treatment as the backup treatment option. Most of the comments in this section are addressed to these alternatives, but the chapter will conclude with comments on other treatment options that DOE should consider.

The first charge is addressed through a discussion in this section of the following four questions:

1.   Were a reasonably complete set of treatment alternatives identified by DOE?
2.   Was the methodology used to screen and rank the treatment alternatives technically sound, and did it lead to the selection of appropriate primary and backup treatment options?
3.   Are the primary and backup treatment options likely to work as described and produce acceptable waste forms?
4.   What other treatment options should DOE consider?

Several of the consultants provided comments that were helpful in responding to these questions, most notably Joseph Byrd, Robert Dillon, Harry Harmon, and Paul Shewmon. Their reports are provided in Appendix D.

The answer to the first question—were a reasonably complete set of treatment options identified?—is a qualified "yes." The Task Team used a somewhat ad hoc approach to identify the set of alternative treatment options in that it relied on the knowledge of Task Team members, two DOE documents (DOE, 1996b, c), and a solicitation to DOE laboratories to identify alternative treatment options. There was no effort made to perform a systematic search for treatment technologies in use in other countries or industries that might be applied to aluminum spent fuel. Nevertheless, the Task Team appears to have succeeded in identifying a reasonably complete set of alternatives, and neither the Principal Investigator (P.I.) nor the consultants invited to the second information-gathering meeting were able to identify other treatment options that should have been considered by the Task Team. Additionally, a query by the P.I. to International Atomic Energy Agency staff did not uncover any other known treatment options.

The affirmative answer to this question is qualified for two reasons. First, the Task Team may incorrectly have eliminated chloride volatility treatment during the initial screening stage. The reason given for eliminating this treatment option that "no experimental work has been completed" (Task Team, 1996, p. 32). In fact, considerable experimental work on chloride volatility was completed between 1950 and 1965 at Argonne National Laboratory (e.g., Jonke, 1965). Although the primary fuel investigated was zirconium based, there were pilot-scale runs on aluminum fuel assemblies and some experimental runs on irradiated fuel. The P.I. did not attempt to determine whether this treatment option would have ranked above the other advanced treatment alternatives had it been evaluated further by the Task Team, but the reason for eliminating it seems to have been based on a lack of awareness of the earlier work.

Second, the Task Team did not consider options for treating the depleted uranium spent fuel, which comprises about 40 percent of the inventory of aluminum spent fuel expected to be received at Savannah River.[19] As noted earlier in this chapter, the Task Team identified this fuel as a candidate for reprocessing because the fuel is declad and is comprised of uranium metal that may not be suitable for disposal in the repository. If reprocessing is not possible for policy reasons (see Chapter 5), then there is no obvious treatment and disposition pathway for this fuel.

The answer to the second question—was the methodology used to screen and rank the treatment alternatives technically sound and did it lead to the selection of appropriate primary and backup treatment options?—also is a qualified "yes." The Task Team developed sufficient

---

[19] The blanket fuel is from the Experimental Breeder Reactor and comprises about 16.6 MTHM (Task Team, 1996, Table 5.2-1) of the 42 MTHM of aluminum spent fuel expected to be received under this program.

technical information and understanding about the treatment alternatives to make reasonable assessments and comparisons of likely performance. Information did not exist, however, to enable the Task Team to determine whether some of the advanced treatment options such as glass material oxidation and dissolution treatment and plasma arc treatment could be made to work in remote handling environments, but this uncertainty appears to have been at least partially reflected in the lower scores and rankings (Table 2.2) for these options.

The Task Team appears to have developed the technical requirements for the alternative options in general terms, but neither the team nor other parts of DOE have developed a complete set of process requirements that would allow an adequate assessment of the options to be made. The most significant set of incomplete process requirements involves repository waste acceptance criteria, which are discussed in more detail in Chapter 3. Most significantly, the acceptability of HEU for disposal in the repository is uncertain, primarily because of criticality and proliferation concerns. If this fuel is not acceptable for disposal, then direct co-disposal treatment should not be selected. The responsibility for establishing repository requirements lies with DOE's Office of Civilian Radioactive Waste Management, not with DOE-Savannah River, and this divided responsibility makes it very difficult to develop an efficient and cost-effective program. Nevertheless, until such requirements are known, implementation of appropriate treatment alternatives cannot proceed without significant financial and schedule risks.

The technical approach used by the Task Team to evaluate and rank the alternative treatment options was appropriate, given their degree of technical maturity and the amount and quality of available data, and the ranking methodology was adequate for screening purposes. The Task Team was consistent in its application of the ranking methodology to the various treatment options, although the various treatment alternatives were at different stages of maturity (e.g., the direct co-disposal treatment technology is more mature than the plasma arc treatment technology) and therefore had different amounts of information available on which to base screening assessments.

Although the Task Team used a quantitative methodology to rank treatment options, the methodology relied to a very great extent on expert judgment of the Task Team members in assigning scaling factors to each of the four evaluation criteria (Table 2.1). Thus, the results are only as good as the judgments being made, a fact noted by the Task Team in its report. Although the P.I. was not personally acquainted with all of the Task Team members before undertaking this project, he has reviewed their backgrounds and believes that team members have the appropriate mix of expertise and experience.

The output of the ranking methodology appears to be consistent with what one would expect. The technologies that scored most highly on the four evaluation criteria tended to be technically mature and relatively simple to implement. As a consequence, estimated development costs were lower for the primary treatment options, and the likelihood of timely implementation was correspondingly higher.

The third question posed above—are the primary and backup treatment options likely to work as described and produce acceptable waste forms?—has two parts: (1) Are the primary and backup treatment options likely to work as described? and (2) Are the primary and backup treatment options likely to produce acceptable waste forms? The answer to the second part of the question is deferred to the next chapter, and the P.I. will focus only on the first part in the following discussion.

Direct co-disposal treatment is technically simple and straightforward to implement. The technologies for drying the fuel, placing it in a container, and sealing the container are readily available and likely can be adapted to this application with little additional development work. Although the composition of the co-disposal container has not yet been determined by DOE, the briefings received at the second information-gathering meeting indicate that it probably will be an austenitic stainless steel. In an atmosphere of dried air at room temperature, the aluminum fuel will form a protective oxide layer, and very little reaction of the waste form with the atmosphere or the stainless steel container for decades of interim storage would be expected.[20]

Melt and dilute treatment is more demanding technically than direct co-disposal treatment and will require a more significant infrastructure, including hot cell space, a melter, and an off-gas treatment system. The radioactive fuel must be melted at temperatures up to about 1000 °C, which will result in the release of gaseous fission products such as krypton and at least partial release of volatile elements such as cesium and iodine. These non-gaseous volatiles must be recovered by an offgas system and recycled into the waste stream or otherwise disposed of. To the P.I.'s knowledge, melt and dilute treatment has not been applied to any other type of spent nuclear fuel. However, all of the technologies needed to make this system function successfully have been used in other applications, and it should be a relatively straightforward exercise to bring them together for aluminum spent fuel treatment. For example, melting and casting technologies have been used for several decades to manufacture unirradiated aluminum fuel elements and irradiated uranium-alloy fuel elements, and commercially available furnaces and ancillary processing equipment can be modified for remote operation. Similarly, offgas systems are in use in several other applications and have even been utilized successfully to capture fission products from the pyrometallurgical processing of the highly radioactive spent fuel from the Experimental Breeder Reactor (ANL, 1963; see Figure 2.4). Although such systems cannot be used directly for aluminum spent fuel treatment, they should be readily adaptable for treating aluminum spent fuel, much of which has been in storage for more than a decade and has lost to decay much of its gaseous radionuclide inventory.

Melt and dilute treatment is worth pursuing despite the additional development and infrastructure requirements because it allows more control over waste form composition and performance characteristics. For example, the waste form can be designed for criticality safety without the

---

[20] See the report of consultant Paul Shewmon in Appendix D.

use of neutron poisons. Additionally, this treatment option would reduce significantly the need for spent fuel characterization and the number of canisters to be interim stored and eventually shipped to the repository.[21]

There is not enough information at present about any of the advanced treatment technologies to select a backup option. In particular, not enough is known at present about electrometallurgical treatment, which was selected as the backup option by the Task Team, to determine whether it will work as described, and additional development work will have to be done to determine the feasibility of applying this treatment technology to aluminum spent fuel.[22] The uranium removal process has been demonstrated successfully on uranium metal-based spent nuclear fuel, but the electrorefining step for removal of aluminum requires further development and testing. In addition, more work must be done to demonstrate that all of the waste streams from this process either can be recycled or will be acceptable for disposal in a repository. These questions are currently being addressed by staff at Argonne National Laboratory. If they are successful and the product uranium is almost free of fission products, the process could face the same proliferation concerns as conventional reprocessing.

The answer to the last question—are there other treatment options DOE should consider?—is "yes." DOE should have given more careful

---

[21] The volume reduction is obtained by the elimination of air spaces in the fuel elements during melting.

[22] Another National Research Council committee has been following the development of electrometallurgical processing for several years and has published several reports (e.g., National Research Council, 1995). This committee has not examined the applicability of electrometallurgical processing to aluminum spent fuel. Argonne National Laboratory staff provided written information on the electrometallurgical process at the second information-gathering meeting of this study.

FIGURE 2.4 (A) A view inside a hot cell at Argonne West showing two remotely operated furnaces for melting highly radioactive fuel. Each furnace (see Figure 2.4B) is covered by a bell jar offgas system (cylindrical metal can) that is lowered over the furnace during melting to capture the gaseous fission products. SOURCE: Argonne National Laboratory.

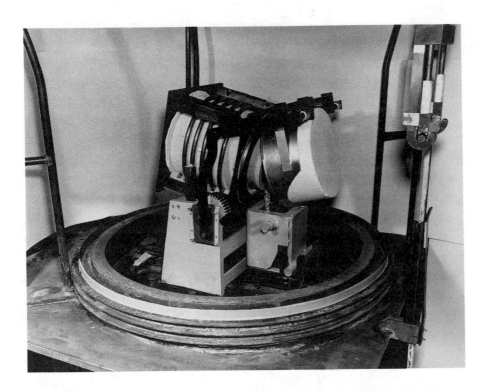

FIGURE 2.4 (B) Closeup photo of one of the furnaces that demonstrates the process for casting a 10 kilogram ingot of highly radioactive fuel. The insulated crucible is heated by an induction coil shown here in the pouring position. The photo shows actual in-cell equipment but the pouring is simulated because of photographic problems with high radiation fields. After casting, the ingot was fabricated into fuel elements for return to the reactor. These furnaces were operated for five years in the mid-1960s as part of the pyrometallurgical fuel cycle used to recycle HEU fuel from the Experimental Breeder Reactor II. SOURCE: Argonne National Laboratory.

consideration to conventional reprocessing of aluminum spent fuel.[23, 24] There appear to be several technical advantages to this option over the others considered by the Task Team. Conventional reprocessing has been demonstrated to work for aluminum spent fuel from production reactors, the costs and risks are well known, the necessary facilities (the Canyons and the DWPF) are currently in operation at Savannah River, and the waste form (borosilicate glass) is assumed to be acceptable for disposal in a repository.

As discussed in more detail in Chapter 5, conventional reprocessing of even a portion of the aluminum spent fuel could significantly reduce the overall costs of treating the total aluminum spent fuel inventory by alleviating the need for additional spent fuel storage space at Savannah River and eliminating the problems with odd-sized fuel elements that may be difficult to process by other methods. There is still some uncertainty about whether the Canyons will be available after 2002 for reprocessing, and future developments in DOE proliferation policy[25] may preclude the use of the conventional reprocessing option except in some special cases (e.g., disposal of damaged spent fuel elements; see Chapter 5 for a fuller discussion of this issue).

It is recommended that DOE-Savannah River undertake a common-basis cost and performance comparison of the two primary treatment alternatives (direct co-disposal and melt and dilute treatment) and conventional reprocessing as part of its process for evaluating and selecting a treatment option.

---

[23] These comments are directed at DOE generally and not to the Task Team, because the Task Team was specifically directed by DOE to look at alternatives to conventional reprocessing.

[24] See also the comments of consultant Harry Harmon in Appendix D.

[25] As noted previously, DOE is expected to release a study on the proliferation implications of conventional reprocessing of aluminum spent fuel from foreign research reactors in the second quarter of FY 98.

3

WASTE-PACKAGE PERFORMANCE CRITERIA

The second charge of the statement of task involves the examination of the waste-package performance criteria developed by the Department of Energy (DOE) to meet anticipated waste acceptance criteria for disposal of aluminum spent fuel and to identify other factors that DOE might consider. In the context of this charge, the term *waste package* refers to the "road-ready" package discussed in Chapter 1. It consists of a metal *disposable canister*[1] that contains several aluminum spent fuel elements or the treated equivalents (e.g., metal ingots produced by melt and dilute treatment). As explained in more detail later in this chapter, the *waste-package performance criteria* comprise the physical, chemical, and thermal characteristics that this waste package must meet to be acceptable for shipment to and emplacement in a repository container. Waste package performance criteria are being developed by DOE-Savannah River staff, and they are based on *waste acceptance criteria* (WAC) that are being created by another part of DOE as part of the repository development program. This program is discussed in more detail in the following section.

This chapter provides a short review of the waste-package performance criteria that are under development by DOE-Savannah River to meet anticipated repository WAC. The chapter also provides comments on some of the technical work being carried out by or under the direction of DOE-Savannah River to demonstrate conformance with these WAC.

---

[1] A stainless steel canister whose primary purpose is to protect the spent fuel or the treated equivalent during interim storage, shipping, and handling operations (see Figure 2.3).

Several sources of information were used to develop this chapter. The main sources of information are the presentations made at the two information-gathering meetings (Chapter 1 and Appendix B) and the following documents:

• Chapter 10, Part 60 of the Code of Federal Regulations (10 CFR 60), *Disposal of High-Level Radioactive Wastes in Geologic Repositories* (USNRC, 1997);
• *Mined Geological Disposal System Waste Acceptance Criteria,* published by the Management and Operating Contractor for the DOE-Office of Civilian Radioactive Waste Management (TRW, 1997a);
• *Alternative Aluminum Spent Nuclear Fuel Treatment Technology Development Status Report* (WSRC, 1997a);
• *Total System Performance Assessment Sensitivity Analysis of U.S. Department of Energy Spent Nuclear Fuel* (CRWMS, 1997a);
• *Acceptance Criteria for Interim Dry Storage of Aluminum-Alloy Clad Spent Nuclear Fuels* (Sindelar and others, 1996);
• *OCRWM[2] Data Needs for DOE Spent Nuclear Fuel* (TRW, 1997b).

BACKGROUND

The Nuclear Waste Policy Act as amended[3] designates Yucca Mountain, Nevada,[4] as the candidate site for a spent fuel and high-level waste repository. Several federal agencies have responsibilities under this

---

[2] Office of Civilian Radioactive Waste Management.
[3] Public Law 97-425 (1982) as amended by P.L. 100-203 (1987) and P.L. 102-486 (the Energy Policy Act of 1992).
[4] Yucca Mountain is located in southern Nevada adjacent to the Nevada Test Site. The candidate repository would be constructed in a bedded tuff several hundred feet above the ground water table. The repository is being designed to contain 70,000 MTHM of spent fuel and vitrified high-level waste from reprocessing.

act for ensuring the safe disposal of spent fuel and high-level waste. The administrator of the U.S. Environmental Protection Agency (EPA) is responsible for promulgating standards for protection of the general environment and the public from releases of radioactive materials from the repository. The U.S. Nuclear Regulatory Commission (USNRC) has the responsibility for establishing the technical requirements that it will use in evaluating applications for authorization to construct a repository, receive and emplace spent fuel or high-level waste, and close and decommission the facility once waste emplacement is completed. The act specifies that these technical requirements are "not to be inconsistent" with EPA standards.[5]

The responsibility for developing and operating a repository at Yucca Mountain lies with DOE.[6] To meet its obligations under the act, DOE and its management and operating contractor[7] are carrying out a detailed characterization program at Yucca Mountain to determine whether the site is suitable for a repository. The program includes geological, geochemical, and engineering studies of the site, including performance assessment (PA) studies to obtain bounding estimates of the long-term behavior of radioactive waste emplaced in the repository. DOE-

---

[5] Standards and regulations were developed for Yucca Mountain during the 1980s by the EPA (in 40 CFR 191) and the USNRC (in 10 CFR 60). But certain provisions of the EPA standards were remanded by judicial action, and the EPA was subsequently directed by the Congress (in the Energy Policy Act of 1992) to develop a separate set of standards for Yucca Mountain. These standards have not yet been issued. Thus, at present, DOE is faced with the difficulty of developing repository and waste package designs to conform to EPA standards that do not yet exist.

[6] Specifically, the act designates the director of DOE's Office of Civilian Radioactive Waste Management as the party responsible for carrying out these tasks, acting under the general supervision of the Secretary of Energy.

[7] In subsequent discussions, the DOE Office of Civilian Radioactive Waste Management and its management and operating contractor will be collectively referred to as DOE-Yucca Mountain.

Yucca Mountain also is developing repository and waste-package designs[8] and detailed WAC for radioactive wastes to be emplaced in the repository.

The WAC comprise the physical, chemical, and thermal characteristics that spent fuel, high-level waste, and associated disposable canisters must conform to for disposal in the repository. These criteria cover the characteristics of the waste and associated disposable canisters that affect the safety of the "back end" of the disposal process—that is, receipt of the waste at Yucca Mountain, transfer of the waste into disposal containers, transport of the waste into the repository, emplacement of waste in the repository drifts, and repository performance. The criteria are based on USNRC regulations (e.g., 10 CFR 60) as well as requirements imposed by DOE-Yucca Mountain to protect worker and public health during repository operations and to meet long-term repository performance requirements.

Before DOE-Savannah River can ship its inventory of aluminum spent fuel to Yucca Mountain, it must obtain a certification from DOE-Yucca Mountain that its wastes conform with the WAC. To this end, DOE-Savannah River is working with DOE-Yucca Mountain to determine which WAC are likely to apply to aluminum spent fuel and what data are likely to be needed to demonstrate conformance. The certification process will involve the submission of an application that contains these required data by DOE-Savannah River to DOE-Yucca Mountain for review and approval. The format and prescribed information to be included in the application have yet to be determined.

---

[8] Specifically, DOE-Yucca Mountain is designing *disposal containers*, which may consist of two or more corrosion-resistant metallic layers and are expected to maintain their integrity for thousands of years. All spent fuel and high-level waste will be placed in one of these containers before being emplaced in the repository, and each container is designed to hold a number of spent fuel assemblies or high-level waste glass logs (see Figure 2.3).

WASTE ACCEPTANCE CRITERIA

The waste-acceptance criteria developed by DOE-Yucca Mountain are provided in the document *Mined Geological Disposal System Waste Acceptance Criteria* (TRW, 1997a)—that was made available for this review. This document provides the WAC for the following types of waste:

1. *Intact Spent Fuel.* Spent nuclear fuel in which cladding integrity is maintained and none of the structural components have been compromised.
2. *Spent Fuel in Disposable Canisters.* Spent nuclear fuel that lacks structural integrity or has fuel cladding degradation that could adversely affect repository performance and therefore must be delivered to the repository in a disposable canister.
3. *High-Level Waste.* Waste generated from reprocessing activities that has been vitrified in borosilicate glass logs.
4. *Other Radioactive Waste.* Waste that is not spent nuclear fuel or high-level waste.

The aluminum spent fuel must conform to the criteria in the second category, because DOE-Savannah River has decided to place the waste in disposable canisters that do not need to be reopened before emplacement in the repository outer container that is being designed by DOE-Yucca Mountain (e.g., WSRC, 1997a, p. 3.3). A list of the applicable WAC for spent fuel in disposable canisters is given in Table 3.1. These are grouped into the following categories:

- general or descriptive criteria;
- physical or dimensional criteria for canistered waste;
- chemical compatibility for canistered waste; and
- thermal, radiation, and pressurization criteria.

TABLE 3.1 Waste Acceptance Criteria for SNF in Disposable Canisters

Criterion No.	Title

General and Descriptive Criteria for Non-Intact SNF

2.1.1	Compliance with Nuclear Waste Policy Act Definition of SNF
2.1.2	Minimum Cooling Time Since Reactor Discharge
2.1.3	Provision that SNF be a Solid
2.1.4	Provision that Wastes Other than Intact SNF be Canistered
2.1.4.1	Canistering of Degraded or Damaged SNF
2.1.4.2	Canistering of SNF Debris and Corrosion Products
2.1.4.3	Canistering of Non-Fuel Components
2.1.5-2.1.19	Placeholder for Future Text

General and Descriptive Criteria for Canistered Waste

2.1.20	Provisions for Disposable Canister Materials
2.1.21	Requirement that Canisters be Sealed
2.1.22	Limits on Free Liquids in Canistered SNF
2.1.23	Maximum Allowable Quantity of Particulates
2.1.24	Limits on Pyrophoric Materials
2.1.25	Limits on Combustible, Explosive, or Chemically Reactive Waste Forms
2.1.26	Provision for Unique, Permanent Canister Labeling
2.1.27	Provision for Tamper-Indicating Devices (TID) on Canisters Not Seal-Welded
2.1.28	Physical Condition Of Disposable Canisters

Physical and Dimensional Criteria for Canistered Waste

2.2.20.1	Dimensional Envelope for Disposable Single-Element Canisters
2.2.20.2	Dimensional Envelope for Disposable Multi-Element Canisters
2.2.20.3	Dimensional Envelope for Disposable DOE-Owned SNF
2.2.21.1	Weight of Disposable Single-Element Canisters
2.2.21.2	Weight of Disposable Multi-Element Disposable SNF Canisters
2.2.21.3	Weight of Disposable DOE-Owned SNF Canisters
2.2.22.1	Capability to Lift Commercial SNF Canisters
2.2.22.2	Capability to Lift DOE-Owned SNF Canisters

Chemical Compatibility Criteria for Canistered Waste

2.3.20.1	Limits on Radionuclide Inventories in Single-Element Canisters
2.3.20.2	Limits on Radionuclide Inventories in Multi-Element Canisters
2.3.21	Limits on Total Fissile Material in a Disposable Canister
2.3.22	Limits on Disposable Canister Criticality Potential
2.3.23	Limits on Organic Materials in Canistered SNF

Thermal, Radiation and Pressurization Criteria for Canistered Waste	
2.4.20	Limits on Total Thermal Output for Disposable Canisters
2.4.21	Limits on Disposable Multi-Element Canister Thermal Design
2.4.22	Limits on Disposable Canister Surface Contamination
2.4.23	Provisions for Canister Internal Pressure
2.4.24	Limits on Disposable Canister Leak Rates

SOURCE: TRW (1997a).

The first two groups of criteria address basic definitions (e.g., what spent fuel is), requirements (e.g., spent fuel cooling time after discharge from a reactor), and disposable canister characteristics (e.g., physical dimensions, weight, construction). These criteria are fairly straightforward and should be easy to conform to through careful documentation and design. The criteria in the last two groups are potentially problematical, because they will require data collection and analysis to demonstrate conformance and because the criteria are uncertain at present and subject to change in the future. Several of these criteria are discussed in more detail in a later section.

OTHER REQUIREMENTS

Although the focus of this chapter is on waste acceptance criteria, it is important to recognize that at least three other sets of requirements must be satisfied before the aluminum spent fuel or its processed equivalent can be shipped to Yucca Mountain for disposal. First, after treatment, the aluminum spent fuel will be placed in interim dry storage at Savannah River until it can be shipped to a repository, and it must conform to a set of interim storage criteria being developed by DOE-Savannah River. For planning purposes, DOE-Savannah River is assuming that interim dry storage will last for up to 50 years, and it is in the process of establishing criteria (Sindelar and others; 1996, WSRC,

1997a) for interim storage that will set limits on fuel corrosion, deformation, cladding integrity, and fission-product release. A summary of these criteria is given in Table 3.2. These criteria were established to minimize fuel corrosion and to maintain criticality control and fuel handleability during the interim storage period.

TABLE 3.2   Criteria for Interim Storage of Aluminum Spent Fuel at Savannah River

Criterion No.	Description
1	Free water remaining within the sealed storage canister after drying is limited to maintain the hydrogen content less than 4% by volume.
2	The lag storage, treatment, and canister storage environments shall limit general corrosion or pitting corrosion to less than 0.0076 cm (0.003 in.) in depth in SNF cladding or in exposed fuel material.
3	The canister storage environment shall preclude the plastic deformation of SNF elements to less than 2.54 cm (1.0 in.) over a fuel assembly length of 91.44 cm (3.0 ft.) and deformation not to exceed 75% of the clearance space between the fuel assembly and storage grid throughout the period of storage.
4	The interim storage environments shall prevent rupture of the SNF cladding due to creep or due to severe embrittlement.
5	Canisters shall be backfilled with helium to 1.5 times atmospheric pressure at room temperature.
6	The storage facility shall be capable of handling canisters from 10 to 15 ft. in length.
7	The interim storage environment shall prevent the SNF cladding temperatures from exceeding 200 °C.

SOURCE: WSRC (1997a).

Second, the road-ready packages must meet U.S. Department of Transportation and USNRC regulations that govern the shipment of nuclear materials. These regulations, which are provided in 10 CFR 71, include requirements for transport configurations, criticality evaluations, accident testing, and quality assurance plans. A detailed consideration of these requirements is outside the current statement of task.

Third, the repository itself must meet certain USNRC radionuclide release limits given in 10 CFR 60[9] and yet-to-be-established EPA dose and possibly ground water standards.[10] To ensure that the repository will comply with these standards, DOE-Yucca Mountain is modeling the long-term performance of the waste forms to be emplaced in the repository, including the aluminum spent fuel waste forms,[11] through its PA program.[12]

---

[9] 10 CFR 60 requires in part that the wastes be emplaced in packages that provide substantially complete containment for 300 to 1,000 years, and that releases from the repository be less than 1 part in 100,000 parts after 1,000 years.

[10] As noted in footnote 5 in this chapter, the EPA will be releasing safety standards for Yucca Mountain in the future. At the present time it is very uncertain what those standards will contain.

[11] That is, the physical and chemical form of the disposal product. For direct co-disposal treatment, the waste form consists of aluminum spent fuel. For melt and dilute treatment, the waste form consists of uranium-aluminum alloy ingots. For conventional reprocessing treatment, the waste form consists of vitrified glass logs.

[12] Currently, the PAs for Yucca Mountain are primarily parametric analyses. It is not clear to the P.I. whether the PA will have a controlling role in repository licensing or, like probabilistic risk assessment in reactor licensing, it will have a minor role, especially given that the data base for PA is so much smaller than the data base for reactor analysis. Irrespective of its use in repository licensing, the PA will play an important role in establishing WAC for the repository, and one would hope for a strong and transparent relationship between the WAC, transportation requirements, and performance of the repository.

RESPONSE TO SECOND CHARGE IN STATEMENT OF TASK

As noted at the beginning of this chapter, the second charge in the statement of task involves an examination of the waste-package performance criteria being developed by DOE-Savannah River for aluminum spent fuel and the identification of other criteria that should be considered. Most of the comments in this section will be addressed to the performance criteria that have been developed by DOE-Savannah River in response to the WAC published by DOE-Yucca Mountain (TRW, 1997a), but additional comments will be made about other criteria that could have a significant impact on the selection of a treatment option.

To address the second charge, answers are provided to the following three questions:

1. Have all of the important waste-package performance criteria been identified by DOE-Savannah River?

2. Are there other performance criteria that should be considered?

3. Is the work under way by DOE-Savannah River appropriate to demonstrate conformance with the various criteria and requirements?

Consultants Francis Alcorn, Robert Bernero, and Valerie Putman provided comments that were helpful in responding to these questions. Their reports are provided in Appendix D.

The answer to the first question concerning the identification of important waste-package criteria (Table 3.1) is a qualified "yes." DOE-Savannah River staff appear to be working closely with their counterparts at DOE-Yucca Mountain to ensure that the important WAC have been identified and that the right kind of work is being done to demonstrate conformance. Perhaps the best evidence of this close working relationship is the analysis being done on nuclear criticality by DOE-Yucca Mountain under a contract from DOE-Savannah River. DOE-Savannah River also appears to have access to draft documents being prepared by DOE-Yucca Mountain that could affect the acceptability of aluminum spent fuel for disposal at the repository. The answer is qualified because, as noted earlier in this chapter, many of the WAC are preliminary and could change significantly as DOE-Yucca Mountain refines the waste package and repository designs. Thus, a continuance of the ongoing dialogue between DOE-Savannah River and DOE-Yucca Mountain will be essential to track and respond effectively to any future changes.

The answer to the second question—should other criteria be considered?—has three parts: (1) for the WAC, (2) for the interim storage criteria, and (3) for the transportation criteria. The answer to the first part of the second question is "no." The current WAC for the candidate repository at Yucca Mountain are very clearly laid out (TRW, 1997a), and the information received during this study from DOE-Savannah River (e.g., WSRC, 1997a) indicates that all of the potentially applicable WAC have been identified and are being addressed through ongoing work. As noted above, however, many of the WAC are preliminary and could change significantly as waste package and repository designs are refined by DOE-Yucca Mountain. Again, a continuation of the dialogue between DOE-Yucca Mountain and DOE-Savannah River will be essential to track and respond effectively to any future changes.

The answer to the second part of the second question—are there other criteria that should be considered for interim storage (Table 3.2) in addition to those that are required for repository acceptance?—is "no." Most of the criteria seem reasonable given the current plans that DOE-Savannah River has to store, retrieve, and process (as necessary) the fuel to put it into road-ready form. One of the criteria, however, appears to be unnecessary. Specifically, criterion 3 in Table 3.2, which sets limits for plastic deformation of the aluminum spent fuel in the disposable canister, seems overly restrictive and potentially costly. The justification given for this requirement is that it will "provide for ready removal of the fuel from a canister and handleability of the fuel" (WSRC, 1997a, p. 3.4). It is not clear why DOE-Savannah River would ever want to remove spent fuel from a disposable canister under normal operating conditions. If the fuel was treated properly before placement in the canister there should be no need to retrieve it prior to shipment to the repository. Even under "abnormal" conditions such as a tipover accident the canister could be sectioned to remove the spent fuel. DOE-Savannah River is encouraged to reexamine the cost and potential benefit of this criterion in view of the unlikely need for future fuel removal.

The third part of the second question—are there other transportation criteria that should be considered?—cannot be answered at this time. Based on discussions with DOE-Savannah River staff, relatively little work has been done to date on establishing criteria to meet transportation requirements. The transportation requirements given in 10 CFR 71 could affect the design of the disposable canister into which the spent fuel or its processed equivalent will be placed for shipment to and emplacement in the repository. DOE-Savannah River should not encounter any significant problems meeting the requirements in 10 CFR 71—highly enriched uranium (HEU) spent fuel is shipped across the country and around the world routinely, and the aluminum research reactor spent fuel now stored at Savannah River was shipped from offsite at some time in the past. DOE-Savannah River must review the shipping requirements before it finalizes the design of its disposable canisters.

The answer to the third question—is the work under way by DOE-Savannah River appropriate to demonstrate conformance?—is a very qualified "yes." Based on the presentations received at the two information-gathering meetings, the documents reviewed, the tour of the Savannah River Technology Center where much of this work is being done, and discussions with the personnel there, the development program under way to demonstrate conformance with the WAC appears to be properly focused and appropriate to the task. This answer is qualified, however, because the short schedule for this project did not allow an in-depth review of all of the ongoing work in the aluminum spent fuel program, and because the WAC on which current activities at Savannah River are based may change once EPA standards and USNRC regulations are issued.

The remainder of this section provides a few comments on WAC developed by DOE-Yucca Mountain that are potentially limiting for treatment option selection either because data collection or analysis will be required to demonstrate conformance or because the criteria are uncertain and subject to change in the future. The following sections provide brief discussion of three such criteria and a discussion of the efforts under way at DOE-Savannah River to document conformance. The criteria are 2.3.20.2 (limits on radionuclide inventories in multi-element canisters), 2.3.21 (limits on total fissile material in a disposable canister), and 2.3.22 (limits on disposable canister criticality potential). A complete list of criteria is provided in Table 3.1.

2.3.20.2: Limits on Radionuclide Inventories in Multi-Element Canisters

Calculated radionuclide inventories in disposable multi-element canisters for a given group of waste shall not average more than the levels listed in [Table 3.3] to be accepted into the MGDS [Mined Geological Disposal System] (all numbers TBV [to be verified]). There are no restrictions on radionuclides excluded from this list. Note: It must be recognized that acceptance limits for radionuclides must be considered preliminary until additional Performance Assessment [PA] analyses are performed, and the repository PA is accepted by the [US]NRC as part of granting a repository operating license. Currently, there is no environmental release standard to which acceptance limits can be set (40 CFR 191 standards are used, even though 40 CFR 191 has been remanded), there are a wide range of assumptions used in the Performance Assessment models that can significantly alter acceptance ranges (these assumptions have varying degrees of concurrence from the [US]NRC), and a number of important mechanisms that will influence acceptance ranges are not yet adequately modeled by the current Performance Assessment codes. Radionuclide inventories footnoted with an asterisk [in Table 3.3] represent those that are the largest contributors to total repository release rates and are, along with their parent radioisotopes, the least likely to have higher acceptance limits in future versions of this document. (TRW, 1997a, p. 4-17–4-18.)

This criterion is of particular concern for direct co-disposal treatment for two reasons.[13] First, DOE-Savannah River will have to characterize the aluminum spent fuel to demonstrate that it meets the limits in the table. Second, the limits shown in the table are subject to change as the PA models are verified or new EPA standards or USNRC regulations are promulgated. A future move to risk-based regulation, for example, could introduce a new set of radionuclide inventory criteria.

The simplest way to demonstrate that spent fuel meets the limits in Table 3.3 is through direct calculation using the fuel property data, which are determined during fuel fabrication, and the reactor operation history to calculate fuel burnup and isotope production. This approach is preferred because it is sufficiently accurate, relatively inexpensive, and technically straightforward. However, the direct calculation approach may be problematical for some of the foreign research reactor fuel and the

---

[13] Waste form characterization is not a problem for melt and dilute treatment or conventional reprocessing, because the physical and chemical properties of the waste form can be controlled during processing.

Research Reactor Aluminum Spent Fuel

TABLE 3.3 Limits on Radionuclide Inventories in Multi-Element Canisters

Isotope[a]	Concentration Average (Ci/waste package)[b]
^{227}Ac	1.79×10^{-4}
^{241}Am	3.73×10^{4}
^{242M}Am	2.16×10^{2}
^{243}Am	2.48×10^{2}
^{14}C*	1.38×10^{1}
^{36}Cl	1.11×10^{-1}
^{244}Cm	1.16×10^{4}
^{245}Cm	3.36
^{246}Cm	6.95×10^{-1}
^{135}Cs*	5.13
^{129}I	3.43×10^{-1}
^{93M}Nb	1.82×10^{1}
^{94}Nb	8.24
^{59}Ni*	2.36×10^{1}
^{63}Ni	3.10×10^{1}
^{237}Np*	4.35
^{231}Pa	3.30×10^{4}
^{210}Pb	6.75×10^{-6}
^{107}Pd*	1.26
^{238}Pu	3.05×10^{4}
^{239}Pu	3.56×10^{4}
^{240}Pu	5.26×10^{3}
^{241}Pu	3.39×10^{5}
^{242}Pu	2.01×10^{1}
^{226}Ra*	2.50×10^{-5}
^{228}Ra	3.10×10^{-9}
^{79}Se*	4.41
^{151}Sm	3.53×10^{3}
^{126}Sn	8.50
^{99}Tc*	1.42×10^{2}
^{229}Th	3.54×10^{-6}
^{230}Th	3.59×10^{-3}
^{232}Th	4.35×10^{-9}
^{233}U	7.01×10^{-1}
^{234}U	1.34×10^{1}
^{235}U	1.68×10^{-1}
^{236}U	2.72
^{238}U	3.07
^{93}Zr	2.38×10^{1}

[a] Isotopes with an asterisk represent the largest contributors to total repository release rates in current versions of the PA.
[b] Curies per waste package.
SOURCE: TRW (1997a).

older domestic research reactor fuel, because the data to perform such calculations may be unavailable, incomplete, or inaccurate. In these cases it may be necessary to use direct measurement techniques such as gamma spectrometry or radiochemical techniques, which can add expense and time to the characterization process. It may also be possible to use bounding analyses based on average values as is done for power reactor fuel.

DOE-Savannah River has initiated a program to evaluate the availability and quality of existing fuel property and reactor operation history data for representative aluminum spent fuel types to determine if such data are adequate to meet existing acceptance requirements (WSRC, 1997a). This activity is necessary, cost effective, and timely. Fuel receipt and storage schedules and costs could be affected significantly by the level and amount of characterization required. The relatively small investment of time required to determine the availability of such data should pay off handsomely in the future in terms of increased throughput and reduced handling as the fuel is received at Savannah River for treatment and interim storage. This activity should continue to receive a high priority.

2.3.21: Limits on Total Fissile Material in a Disposable Canister

There are no limits on total fissile material in disposal canisters based on current waste-package and performance assessment analyses. Note: Analyses to establish limits on total fissile material are ongoing but were not available in time for inclusion in this version of the MGDS WAC. Until such limits are available, this text represents a placeholder for a future quantified criteria. (TRW, 1997a, p. 4-20.)

DOE-Yucca Mountain has placed no limits on the amount of fissile material[14] in waste packages, but the guidance language provided above suggests that it intends to establish such limits in the future. Limits on fissile material could be imposed either as a matter of policy, to reduce the potential for criticality, or to limit the release of radionuclides from the repository to meet regulatory or safeguard standards.

Enriched aluminum spent fuel contains less fissile material than typical commercial spent fuel,[15] so fissile material limits are not likely to be a problem if DOE-Yucca Mountain sets a single limit for all spent fuel to be disposed in the repository. If DOE-Yucca Mountain determines that lower fissile material limits are required for aluminum spent fuel, or if lower limits are required for all HEU fuel, then disposal of such fuel using direct co-disposal treatment may not be possible.

2.3.22: Limits on Disposable Canister Criticality Potential

Canistered SNF [spent nuclear fuel] *entering the MGDS shall be shown to have a calculated k_{eff} of 0.95 or less, after allowance for bias in calculation methods and uncertainty in the empirical data used to validate the method of calculation assuming the following conditions:*

   • *All canister basket structure (other than components made from titanium, zircaloy, or other*

---

[14] The important fissile isotope in enriched aluminum spent fuel is uranium-235 (^{235}U). The important fissile isotopes in commercial spent fuel are ^{235}U and plutonium-239 (^{239}Pu).

[15] A canister of commercial spent fuel will contain about 80 kg of ^{235}U and 60 to 70 kg of ^{239}Pu + ^{241}Pu (Oak Ridge Light Water Reactor Radiological Database). A disposal container of HEU aluminum spent fuel will contain a maximum of about 30 kg of ^{235}U and only small amounts of ^{239}Pu relative to commercial spent fuel (based on data from CRWMS, 1997b).

extremely corrosion-resistant materials) have collapsed and degraded into component corrosion products (e.g., FeO_2 from carbon steel basket materials).

* *All supplemental neutron absorber materials (e.g., boron), except hafnium, have degraded and are no longer part of the waste package.*

* *Assembly hardware has degraded and all fuel assemblies are touching in a optimum reactivity condition (assuming a corrosion resistant zircaloy clad fuel).* [16]

* *SNF reactivity has increased to the peak levels in the early years after reactor discharge.* (TRW, 1997a, p. 4-20.)

One of the most important objectives in designing the repository and waste canister is to prevent the possibility of criticality events [17] from occurring during or after the emplacement of spent fuel. [18] Current USNRC regulations in 10 CFR 60 require DOE-Yucca Mountain to demonstrate with a 5 percent margin of safety [19] that the probability of a criticality event in the repository is less than one in a million (1×10^{-6}) for 10,000 years following disposal. The USNRC is revising this rule and may impose different or additional requirements, for example, that the dose consequences of a criticality event also be determined.

---

[16] This criterion is not relevant to aluminum spent fuel.

[17] A criticality event is a self-sustaining nuclear reaction, much like that in a nuclear reactor. In a repository, such an event can occur when the fissile materials from the spent fuel (e.g., ^{235}U) are brought into certain geometric configurations in the presence of water. A criticality event in the repository would result in the release of energy and neutrons but might not be detectable at the surface.

[18] See the comments of consultants Francis Alcorn and Valerie Putman in Appendix D for more details.

[19] Conventionally expressed as $k_{eff} \le 0.95$.

DOE-Yucca Mountain intends to demonstrate compliance with this regulation through careful attention to repository design, the placement of disposal canisters in the repository, the internal configuration of spent fuel in the waste packages, and possibly limits on fissile materials in the waste packages (criterion 2.3.21). The upper-limit probability of a criticality event for a given set of design criteria can be determined through calculation using worst-case assumptions for waste package degradation, geochemical reactions, and ground water flow. Examples of the types of worst-case assumptions used in the calculations are shown in the bulleted paragraphs in the WAC (see the italicized text above).

The criticality potential criterion will require the most attention for direct co-disposal of aluminum spent fuel because of its ^{235}U enrichment relative to spent commercial fuel. DOE-Savannah River appears to recognize the importance of this criterion and has developed a cooperative program with DOE-Yucca Mountain to establish the technical viability of the direct co-disposal option with respect to criticality. This work is described in WSRC (1997a).

Work is currently under way at DOE-Yucca Mountain to assess the criticality potential of aluminum spent fuel for the direct co-disposal option and is occurring in three phases: Phase 1, which examines the criticality potential of degraded fuel in an intact disposal canister (i.e., the inner canister in Figure 2.3); Phase 2, which examines the criticality potential when the degraded fuel is released from the disposable canister into the co-disposal waste package (the outer container in Figure 2.3); and Phase 3, which examines the criticality potential when the degraded fuel is released from the co-disposal waste package into the repository.

Briefings were received from DOE on phases 1 and 2 of this work during the two information-gathering sessions, and the written report of the phase 1 work, which was completed in mid-1997 (CRWMS, 1997b), was reviewed by the P.I. and consultants. The work to date appears to be technically sound, and neither the P.I. nor the two criticality consultants invited to the second information-gathering meeting (Appendix D) identified any significant major flaws in the design or execution of these analyses.[20]

The phase 1 work suggests that aluminum spent fuel will conform to criterion 2.3.22 if neutron absorbers (e.g., borated stainless steel) are added to the waste packages, and the phase 2 work described in the information-gathering sessions seems to indicate that neutron absorbers also can be used to meet this criterion for the degraded container scenario. Additional work will be required to confirm this result only if direct co-disposal treatment is selected, because the melt and dilute treatment and conventional reprocessing can be designed to produce waste forms that do not contain HEU.

Not enough work has been done on the criticality potential external to the container (i.e., the phase 3 study mentioned previously) to determine whether aluminum spent fuel—especially HEU fuel—will conform to criterion 2.3.22 for direct co-disposal. If it cannot, the direct co-disposal treatment option may be eliminated for all or certain types of aluminum spent fuel. [21]

CONCLUSIONS

DOE-Savannah River appears to have identified all of the significant waste acceptance criteria for aluminum spent fuel and is engaged in the proper process (through close consultation with DOE-Yucca Mountain) to demonstrate conformance. As noted in the foregoing discussion, however, several of the WAC are poorly defined at present or

---

[20] The criticality consultants (Francis Alcorn and Valerie Putman) did note several minor concerns about this work in their reports (Appendix D), but the P.I. assumes that these will be addressed by DOE before the criticality work is completed.

[21] Of course, changes to the design of the repository, including engineered barriers, also could influence the acceptability of HEU spent fuel for disposal.

may be subject to significant future change. In fact, for several reasons including lack of an EPA standard and corresponding USNRC regulations, it may be quite some time before DOE-Savannah River knows with certainty whether direct co-disposal is viable for all fuel types.

The current state of uncertainty has significant implications for the "path forward" for selecting spent fuel treatment options. Three initial conclusions based on these facts are offered below:

1. A single treatment option may not be suitable for all types of aluminum spent fuel.

2. The aluminum spent fuel program will need to maintain flexibility in selecting treatment options until there is more complete information on the WAC and other requirements.

3. A path forward that involves phased decision-making in the selection and implementation of alternative treatment options is indicated.

These conclusions are developed in greater detail in the last chapter of this report.

4

COST AND SCHEDULE

The third and final charge of the statement of task directs the National Research Council to provide, to the extent possible given the accelerated schedule for this project, an assessment of the cost and timing aspects associated with implementation of each aluminum spent fuel treatment option. The four-month schedule for information gathering and report development did not permit an in-depth review of cost and schedule estimates for the alternative treatment options. Instead, the review has been focused on the methodologies used to estimate costs to see if they follow generally accepted practices, are applied consistently, and result in estimates that are useful for comparative and programmatic purposes. Many of the comments in this chapter are focused on the primary treatment options identified by the Task Team (direct co-disposal treatment and melt and dilute treatment), the baseline treatment option (conventional reprocessing), and hybrids of these options, because these appear to be superior to other treatment options identified by the Task Team as noted in Chapter 2.

The main sources of information used in this assessment are the presentations made at the two information-gathering meetings and the following documents:

• *Technical Strategy for the Treatment, Packaging, and Disposal of Aluminum-Based Spent Nuclear Fuel*, Volumes 1 and 2 (Task Team, 1996).
• *Savannah River Site Aluminum-Clad Spent Nuclear Fuel Alternative Cost Study* (WSRC, 1997b).

The first part of this chapter describes the cost and schedule estimates made by the Task Team and by Westinghouse Savannah River staff in the above-referenced reports. The last part of the chapter provides comments on the completeness of this work and its usefulness for comparative and programmatic purposes.

TASK TEAM REPORT COST AND SCHEDULE ESTIMATES

Cost was one of the four ranking criteria used by the Task Team to compare the nine alternative treatment options discussed in Chapter 2. The Task Team referred to the cost estimates that it developed as "conceptual" and useful for comparative purposes only.[1] The basis for the cost estimates included Task Team and third-party judgments, the latter primarily from "advocates" for each of the treatment options, costs for comparable applications, and simple calculations. The Task Team estimated the major cost components of an aluminum spent fuel handling, treatment, storage, and disposal system and, for each of these components, the Task Team developed a consistent set of methodologies for estimating costs. The objective was to achieve consistency in cost estimates for the various treatment options rather than to provide absolute estimates of system costs.

To develop comparable cost estimates, the Task Team made several assumptions about the design and implementation of the treatment technologies, the most important of which are given below.

---

[1] The estimates developed by the Task Team did not include factors such as indirect costs and contingencies, nor did they provide detailed breakdowns of facility and human resource costs.

1. *Schedule.* The Task Team assumed that aluminum spent fuel would be received and treated at Savannah River until the year 2035. The cost estimates covered the handling, treatment and interim storage, and disposal of all of the aluminum spent fuel received by Savannah River during this period.

2. *Facility Use.* The Task Team assumed that the existing infrastructure at the Savannah River site, including existing buildings and secondary waste treatment facilities, would be used in the treatment and storage program whenever practical. For example, the Task Team assumed that the aluminum spent fuel would be received and stored in two existing wet storage facilities, the L-Basin and the Receiving Basin for Offsite Fuels (RBOF; see Figure 1.2). The Task Team also assumed that liquid high-level waste (HLW) streams from conventional reprocessing would be disposed of in the HLW tanks at Savannah River and eventually vitrified in the existing Defense Waste Processing Facility (DWPF). The Task Team noted that the use of existing facilities reduced the estimated costs of implementing most of the treatment technologies. Nevertheless, the Task Team determined that existing facilities were inadequate for all steps in the treatment and storage program and that a new spent fuel transfer facility (for receipt, handling, and packaging of spent fuel) was required for all of the treatment options.

3. *Spent Nuclear Fuel Receipt Schedule.* The Task Team assumed that aluminum spent nuclear fuel receipts at Savannah River would follow the schedule shown in Figure 4.1. This schedule was based in large part on the capacity of existing facilities at SRS to receive and store the spent fuel under existing operating conditions.

4. *Schedules for Implementation of Treatment Options.* The Task Team noted that the cost of each alternative treatment option would depend to a great extent on how quickly the option could be implemented. The Task Team assumed the following startup dates based on subjective judgments of the relative "maturity" of each treatment option:

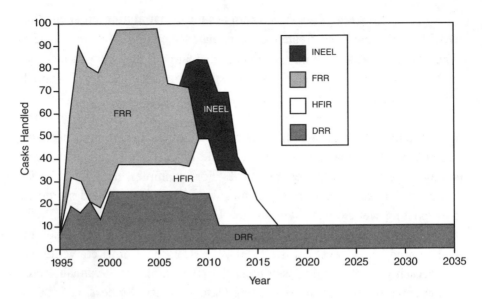

FIGURE 4.1 Projected receipt schedule at Savannah River for aluminum
spent fuel. NOTE: HFIR = High Flux Isotope Reactor spent fuel; FRR =
Foreign Research Reactor spent fuel; INEEL = Idaho National
Engineering and Environmental Laboratory aluminum spent fuel.
SOURCE: Task Team (1996), Figure 4.2-1.

- 2001 startup date for direct disposal or direct co-disposal
treatments.
- 2003 startup date for press and dilute and melt and dilute
treatment.
- 2005 startup date for electrometallurgical treatment.
- 2006 startup date for plasma arc, dissolve and vitrify, or glass
material oxidation and dissolution treatment.

The Task Team noted that these dates were "aggressive" and would require acceleration of budgeting, appropriations, and management practices.

5. *Other Assumptions.* The Task Team made a variety of other assumptions in its cost estimates, two of which are worth noting here. First, the Task Team assumed that the wet basins would be deinventoried as soon as possible after the spent fuel transfer facility was opened to reduce operating costs. Second, the Task Team assumed that the treated fuel would be shipped to the repository beginning in 2020.

The Task Team's conceptual cost estimates are shown in Table 4.1, and a brief explanation of the major cost factors is given in Table 4.2. The last column in Table 4.1, the "cost comparison point," was used by the Task Team as the comparative cost estimate for each treatment option. The cost comparison point was calculated by summing the conceptual costs for each of the cost factors (i.e., the first five cost columns of Table 4.1) and adjusting these for two factors. The first factor, cost "adjustments," involves additional costs or credits associated with each treatment option. For example, the $60 million cost adjustments (i.e., a $60 million expense) for the first four treatment options in Table 4.1 reflect the costs of an additional drying step to condition the fuel for interim storage.[2] For electrometallurgical treatment, the minus $220 million cost adjustment (i.e., a credit of $220 million) reflects the estimated value of the recovered uranium in the commercial fuel market.

The second factor, "relative cost uncertainty," is a measure of the uncertainty of the cost estimates for each treatment technology. The cost uncertainty was estimated by aggregating the uncertainties for each of the major cost factors shown in Table 4.2. These uncertainties were based on

---

[2] The Task Team determined that some of the fuel would have to be hot vacuum-dried to reduce the water available to drive corrosion reactions during interim storage.

TABLE 4.1 Task Team Estimate of Treatment Option Costs

Treatment Option	Conceptual Costs by Category[a]						Relative Cost Uncertainty[c]	Cost Comparison Point
	Storage and Handling	Transfer and Packaging	Treatment	Interim Storage	Disposal[b]	Adjustments		
Direct disposal	280	440	0	120	440	60[d]	20	1,400
Direct co-disposal	280	430	0	130	210	60[d]	70	1,200
Press and dilute (20%)	350	420	230	100	90	60[d]	150	1,400
Press and dilute (2%)	350	440	230	120	200	60[d]	160	1,600
Melt and dilute	350	390	270	100	90	0	150	1,300
Plasma arc	460	380	450	90	90	0	440	1,900
GMODS[e]	460	390	410	110	140	0	410	1,900
Dissolve and vitrify	460	390	720	110	140	0	180	2,000
Electrometallurgical	440	360	600	0	50	-220[f]	400	1,600
Processing and co-disposal	430	170	640	50	90	-180[f]	10	1,200

[a] Cost estimates are shown in millions of dollars and are rounded to the nearest $10 million.

[b] Includes transportation and disposal operations and a prorated share of repository development costs.

[c] Adjustment for relative uncertainty among technology options as discussed in the text.

[d] Estimated costs for a hot vacuum drying facility for fuels that will be direct disposed.

[e] Glass material oxidation and dissolution.

[f] Credit for sale of Uranium-235 on the commercial market.

SOURCE: Task Team (1996), Table 4.2-2.

synoptic judgments by the Task Team of the reliability of the cost data and differences in the technical maturities of the treatment options. In general, the treatment options that are more complex or less mature in a technical sense tend to be associated with higher cost uncertainties. Thus,

TABLE 4.2 Significant Cost Factors in the Aluminum Spent Fuel Dispositioning Program

Cost Factor	Description
Wet storage and handling	Primarily the cost of operating and maintaining existing wet storage facilities at Savannah River (L-Basin and RBOF)
Transfer and packaging	Pre- and post-treatment handling costs, including the cost of spent fuel transfer facility
Treatment	Actual costs associated with treating the waste to put it into a form acceptable for interim storage and repository disposal
Interim storage	The cost of constructing and maintaining an interim storage facility (a modular dry vault) of a size scaled for the number of waste canisters produced by each treatment option
Disposal	Costs for transportation of waste to the repository from Savannah River, placement of wastes in disposal canisters, and emplacement in the repository

SOURCE: Task Team (1996).

for example, direct co-disposal treatment has a lower uncertainty because the technology is relatively mature. Plasma arc treatment is less technically mature and is therefore associated with a significantly greater uncertainty.

A conceptual cost estimate for processing and co-disposal treatment (last row in Table 4.1; see Chapter 2 for a description of this treatment option) was included in the analysis for comparative purposes. The Task Team assumed that aluminum spent fuel at Savannah River would be reprocessed in the H Canyon at Savannah River until 2008, and the aluminum spent fuel received after that date would be treated using direct co-disposal treatment.

ALTERNATIVE COST STUDY

In December 1997, Westinghouse Savannah River Company released a report (WSRC, 1997b, hereafter referred to as the "alternative cost study") that provided life-cycle cost estimates for the period 1998-2037 for the aluminum spent fuel treatment program at Savannah River. This alternative cost study builds on the work in the Task Team report and attempts to provide more realistic cost estimates that can be used for program planning and decision purposes.

In this study, life-cycle costs for each of the treatment technologies were estimated using a "bottoms-up" approach. Estimates for wet storage and handling costs were made using the current operational costs of the L-Basin and RBOF facilities. Estimates for treatment, handling and packaging, and interim storage were made by costing out the required equipment and facility space and by estimating the number of staff and shifts needed to complete the work. Transportation and disposal cost estimates were based on the latest data available from DOE-Yucca Mountain. The cost estimates included indirect costs and contingencies, some financing costs for privatization (see the following paragraph), the costs for U.S. Nuclear Regulatory

Commission [USNRC] licensing[3] and International Atomic Energy Agency [IAEA] safeguards and security controls,[4] and adjustments for inflation.

The cost estimates also reflected several changes in management plans, schedules, and other programmatic assumptions since the Task Team report was published. A detailed discussion of these changes is beyond the scope of the present report, but three significant changes are worth noting. First, the alternative cost study estimates were based on privatization of the aluminum spent fuel treatment program at Savannah River. Costs were adjusted for financing and a five-year capital recovery period to be consistent with Westinghouse Savannah River methods for estimating costs. Second, the estimates were based on a more realistic schedule for implementing the various treatment options. An implementation date of 2006 was assumed for direct co-disposal, press and dilute, and melt and dilute treatment, and a 2011 implementation date was assumed for the other treatment options.

Third, the alternative cost study also considered the use of conventional reprocessing to treat part of the aluminum spent fuel inventory and assumed that shipments of some of the aluminum spent fuel now being stored at the Idaho Engineering and Environmental Laboratory (INEEL) could be accelerated. The alternative cost study provides cost estimates for conventional reprocessing of aluminum spent fuel until about 2010, followed by either direct co-disposal treatment, melt and dilute treatment, or continued reprocessing in a new dedicated facility.

---

[3] DOE is self-regulating and is not required to obtain USNRC licenses for its facilities. If DOE decides to privatize the aluminum spent fuel treatment program, however, the contractor selected to run the program will have to obtain USNRC licenses for its facilities (under 10 CFR 72), even if they are constructed on the Savannah River site.

[4] This involves verification of facility designs, records management, inspections, and containment and surveillance activities carried out in accordance with 10 CFR parts 73, 74, and 75.

From the information made available to the P.I. it was not possible to separate the technical, economic, and policy reasons for proposing the latter plan in which a dedicated facility would be built to continue reprocessing.

The estimates of life-cycle costs adjusted to 1998 dollars are listed in Table 4.3, and the results of a sensitivity analysis of the cost estimates are shown in Figure 4.2. The sensitivity analysis was performed by assuming a +30 percent uncertainty in capital costs, +20 percent on new facility operating costs, +40 percent on repository costs, and +20 percent on uranium values (for resale of recovered uranium to the commercial market). Not surprisingly, the less complex or more mature treatment options (e.g., conventional reprocessing and direct co-disposal treatment) tend to have the lowest estimated costs and smallest cost uncertainties, whereas the more advanced treatment technologies (e.g., plasma arc treatment) are associated with the highest cost estimates and largest uncertainties.

COMPARISON OF TASK TEAM AND
ALTERNATIVE COST STUDY ESTIMATES

The cost estimates provided by the Task Team and the alternative cost study are not in a strict sense directly comparable because they are based on different sets of planning assumptions and are indexed to different budget periods.[5] Nevertheless, three general observations that can be made about these estimates for the purpose of a subsequent discussion of the final charge in the statement of task.

The first observation is that the cost estimates provided in the alternative cost study are significantly higher than the estimates in the Task Team report. These differences range from $830 million for processing and direct co-disposal treatment to about $1.8 billion for

---

[5] The periods are 1996-2035 for the Task Team report and 1998-2037 for the alternative cost study.

TABLE 4.3 Life-Cycle Costs[a] for Aluminum Spent Fuel Treatment Program from Alternative Cost Study

Alternative	Wet Storage and Handling	Transfer, Storage, and Treatment	Fuel and Waste Processing	Repository Disposal	Uranium Credits	Total
Co-disposal	730	1,370	0	170	0	2,270
Melt and dilute	730	1,430	0	50	0	2,210
Press and dilute	730	1,670	0	80	0	2,480
Electrometallurgy	730	2,690	0	30	-270	3,180
Dissolve and vitrify	730	2,840	0	200	0	3,770
GMODS	730	2,470	0	200	0	3,400
Plasma arc	730	2,560	0	80	0	3,370
Reprocess and co-disposal[b]	750	740	670	70	-200	2,030
Reprocess and melt and dilute[c]	750	880	670	30	-200	2,130
Reprocess[d]	750	1,180	670	30	-270	2,360

[a] Cost estimates are shown in millions of 1998 dollars and are rounded to the nearest $10 million.
[b] Reprocessing in the Canyons to last until 2010, followed by direct co-disposal treatment of remaining fuel.
[c] Reprocessing in the Canyons to last until 2010, followed by melt and dilute treatment of remaining fuel.
[d] Reprocessing in the Canyons to last until 2010, followed by reprocessing of the remaining fuel in a new dedicated facility.

SOURCE: WSRC (1997b).

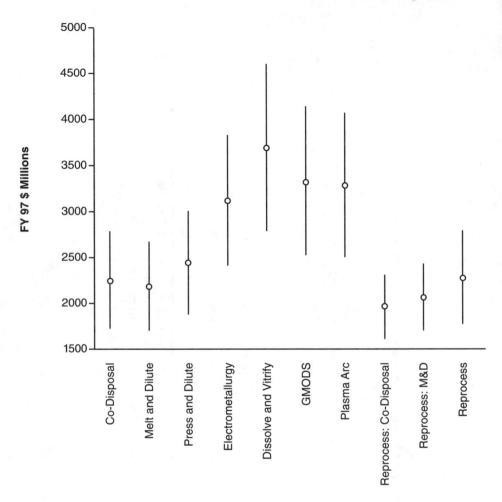

FIGURE 4.2 Sensitivity analyses of the life-cycle cost estimates for the various spent fuel treatment options. Open circle represents the estimated cost, and vertical bars indicate the uncertainty range. NOTE: GMODS = glass material oxidation and dissolution treatment; M&D = melt and dilute treatment. SOURCE: WSRC (1997b).

dissolve and vitrify treatment. The alternative cost study estimates are of total system costs based on more realistic schedules and system requirements.

The second observation is that although the cost estimates in the two studies are significantly different, the relative rankings of the various classes of alternative treatment options are not. In both sets of estimates, the less complex and more mature treatment options are less costly than the more complex treatment options. The least expensive options (processing and direct co-disposal and processing and melt and dilute treatment) are those that rely to the greatest extent on proven treatment technologies (i.e., conventional reprocessing).

A third observation is that there is a relatively small range of estimated costs for the various treatment options, particularly the more mature treatment options. In the Task Team estimate, for example, total system costs for the various treatment options range from $1,200 million to $2,000 million, and costs for the more mature treatment options (direct disposal, direct co-disposal, melt and dilute, press and dilute, and processing and direct co-disposal) range from $1,200 million to $1,600 million. In the alternative cost study, the estimated life-cycle costs of these mature treatment options range from $2,030 million to $2,480 million.

At least three hypotheses can be offered to explain the similarity in costs for the mature treatment options: (1) the cost estimates are incomplete; (2) the costs of the mature treatment options are similar; or (3) the costs of the mature treatment options are different but comprise only a small part of the overall treatment and storage costs. The cost breakdown in the Task Team report (Table 4.1) suggests that the third hypothesis probably is most nearly correct. The estimated treatment costs for the mature technologies are significantly different—they range from zero for direct co-disposal to $270 million for melt and dilute—but account for only a small part of the total system costs for these treatment options. In other words, most of the costs are for handling, storage, packaging, and disposal, not for treatment itself. In the alternative cost study, the distinction between treatment costs and storage and handling costs cannot be made because all of these costs are grouped into a single estimate.

RESPONSE TO THIRD CHARGE IN STATEMENT OF TASK

The final charge of the statement of task involves the assessment of the cost and timing aspects associated with implementation of each spent nuclear fuel treatment option. This charge is addressed through a discussion of the following three questions:

1. Do the cost estimates account for all of the major cost factors in the aluminum spent fuel treatment program?
2. Are the cost and schedule estimates suitable for comparison of the options and selection of one or more preferred alternatives?
3. Are the cost and schedule estimates suitable for budget planning purposes?

Several of the consultants provided comments that were helpful in responding to these questions, most notably Maurice Angvall, Brian Estes, and Richard Smith. Their reports are provided in Appendix D.

The answer to the first question—do the cost estimates account for all of the major cost factors in the aluminum spent fuel treatment program?—is "yes." The major cost factors of the system for receiving, treating, handling, storing, and disposing of aluminum spent fuel for each of the treatment options were identified in the Task Team report, and systematic cost estimates for these major cost factors were developed in the alternative cost study report. The cost estimates were reasonably transparent in both reports: that is, both reports provided reasonably complete cost breakdowns, a list of the programmatic assumptions used in the cost estimates, and an explanation of the methodologies used to estimate uncertainties in total system costs.

Of the two cost studies, the alternative cost study is a more complete estimate of total system costs. This study includes detailed breakdowns of equipment, facilities, and manpower requirements for each treatment option. The cost estimates were constructed using reasonable scaling factors, contingencies, and inflation factors, and they account for IAEA security and safeguard costs and USNRC licensing costs.

The answer to the second question—are cost and schedule estimates suitable for comparison of options and selection of one or more preferred alternatives?—is a qualified "yes." The cost estimates in both the Task Team report and the alternative cost study appear to be sufficiently complete for comparative purposes and for selecting a small number of alternative treatment options for further consideration. The observation noted above that the relative costs among the options in the two reports suggests that the major cost factors have been identified and costs have been estimated adequately in a relative sense.

The schedules laid out in the Task Team report were clearly unrealistic, but this does not appear to have had a significant effect on the selection of treatment options. The schedules laid out in the alternative cost study report are more realistic but still appear to be somewhat ambitious, and there is no provision in the cost estimates for additional program delays. Additional significant program delays could add substantially to the costs for this program.

The answer to the question is qualified because costs did not turn out to be a particularly effective discriminator of the various treatment options, mainly because the treatment options themselves comprised a relatively small part of overall systems costs. There was not much consideration given in either the Task Team report or the alternative cost study reducing overall systems costs by considering alternatives in the fuel receipt schedule shown in Figure 4.1. As noted previously, the fuel receipt schedule used in the Task Team report was based largely on current handling and storage capabilities at Savannah River, and no consideration was given to how changes to this schedule could affect system costs or the selection of alternative treatment technologies. The alternative cost study did examine the impact of accelerating the receipt of aluminum spent fuel from INEEL on processing treatment options, but did not consider other potential flexibilities in the schedule. Additional comments on this point are offered in the next chapter

The answer to the final question—are the cost and schedule estimates suitable for budget planning purposes?—is "no." Although the cost and schedule estimates in the alternative cost study are clearly more realistic than those in the Task Team report, the schedules are still very ambitious and depend to a great extent on the timely completion of work by other parts of DOE. For example, DOE-Savannah River will not be able to select the direct co-disposal treatment option until the acceptability of aluminum spent fuel for direct disposal is established by DOE-Yucca Mountain and the proliferation policy issue (Chapter 5) is resolved. The repository and engineered barrier designs at Yucca Mountain are changing and will continue to do so for at least the next two years, and a definitive PA may not be available until after the environmental impact statement (EIS) and record of decision (ROD) for the aluminum spent fuel program are released.[6] The nonproliferation study currently under way in another part of DOE (Chapter 5) also could significantly impact schedules, budgets, and the selection of treatment options, especially conventional reprocessing.

The cost and schedule estimates also are limited by the lack of conceptual designs for some of the treatment facilities and because some of the process steps have not yet been demonstrated to work for aluminum spent fuel. This affects not only construction and operating costs, but also the decontamination and decommissioning costs of any such facility. Additionally, the alternative cost study assumes privatization of the

---

[6] As noted in Chapter 1, DOE-Savannah River plans to issue the EIS-ROD in 1999.

treatment program, but DOE experiences with cost estimates for privatization have not been very reliable in the past.

The cost estimates also do not consider the impacts of program delays on costs and schedules. Some amount of delay seems inevitable even under the best of circumstances and could come from several quarters. DOE must decide, for example, whether to pursue the project under the privatization program, and if so, it must prepare a solicitation, review bids, and negotiate a contract. A budget for the program must be developed and submitted to the office of Management and Budget (OMB) and to Congress, and funds must be authorized and appropriated. Facilities must be designed and constructed and must pass environmental, health, and safety reviews, including USNRC reviews. Treatment equipment will have to be constructed and tested, and unanticipated problems will have to be addressed.

The program has been unable to meet the schedules outlined in the Task Team report, which was published only two years ago, and a number of additional delays of varying significance will no doubt be encountered as the program moves forward. There is no allowance for such delays in either the Task Team report or the alternative cost study, although DOE-Savannah River staff did express sensitivity to these issues at the information-gathering sessions. DOE-Savannah River will have to incorporate sufficient budgeting and scheduling flexibility into its planning to deal with such delays.

5

CONCLUDING OBSERVATIONS

The primary focus of this report is on options for treating aluminum spent fuel. However, spent fuel treatment is just one component of a much larger and complex *aluminum spent fuel disposal program*, a program that is slated to last for about 40 years and cost in excess of $2 billion. During the course of this study it has become increasingly clear that sound decisions on treatment options cannot be made in isolation of this larger program. This concluding chapter offers some general observations about the overall program and how it impacts the treatment selection process, and also offers some suggestions on how DOE-Savannah River might use this knowledge to make more effective treatment selection decisions.

The aluminum spent fuel disposal program is a complex web of activities at multiple sites around the world, ranging from operations at foreign and domestic research reactors that generate aluminum spent fuel to the repository development program at Yucca Mountain. Several parties have responsibilities for activities that take place in this program, and the decisions made by one party can have significant impacts on costs, schedules, and current or planned operations elsewhere in the program. For the following discussion, it is useful to think of the aluminum spent-fuel disposal program as being comprised of the components shown in Table 5.1. The "front end" of the program involves the generation of spent fuel in the foreign and domestic research reactors, an activity that is expected to continue until at least 2035. The "back end" of the program involves the emplacement of treated spent fuel in the repository and the decontamination and decommissioning (D&D) or reuse of the treatment facilities at Savannah River, activities that are expected to last well into the twenty-first century.

94

An important observation that can be made from inspection of Table 5.1 is that DOE-Savannah River has little if any control over the front and back ends of this program (i.e., components 1, 8, and 9 in Table 5.1) and limited control over components 2 and 7. It does, however, have the responsibility for ensuring that the other components of the program (i.e., components 3-6 in Table 5.1) are compatible with the front and back ends, even though compatibility requirements are not well defined at present. In particular, DOE-Savannah River must select one or more treatment options for aluminum spent fuel that will meet repository waste acceptance criteria, which have yet to be finalized; design treatment and storage facilities that are sized appropriately to waste streams, which are subject to future change; and provide for interim storage of the processed waste until the repository, which is yet to be designed, licensed, or constructed, is able to accept it.

The spent fuel disposal program is a *systems* problem in the classic sense. It involves several interacting components, each associated with different programmatic factors (e.g., cost, time, safety, policy constraints), multiple responsible parties, and different levels of uncertainty (e.g., the right-most column in Table 5.1). The selection of aluminum spent fuel treatment options in the face of such uncertainties calls for a phased strategy in which critical programmatic decisions—that is, decisions that involve major program directions and commitments of funds—are made and implemented when the information needed to base sound choices becomes available. The acquisition of information for decision making also is an important part of the phased-strategy approach, both the acquisition of existing data from third-party sources and the generation of new data to fill information gaps. Of course, the phased strategy recognizes that there may be trade-offs between information acquisition and costs of delayed decisions and seeks to maximize the former and minimize the latter.

In the context of the aluminum spent fuel treatment activities at the Department of Energy's (DOE's) Savannah River site, the primary

TABLE 5.1 Components of Aluminum Spent Fuel Program

Program Component	Responsible Parties	Comments
1. Generation of spent fuel	Reactor owners	Future spent fuel generation will be affected by changes in operating policies, externally imposed regulations, and economic conditions
2. Spent fuel return to SRS	Reactor owners and DOE-Savannah River	Currently limited by number of shipping casks available to aluminum spent fuel program
3. Spent fuel receipt at SRS	DOE-Savannah River	Currently limited by capacities of receiving facilities at Savannah River
4. Spent fuel storage at SRS	DOE-Savannah River	Currently limited by availability of wet basin storage (L–Basin and RBOF) at Savannah River
5. Spent fuel treatment at SRS	DOE-Savannah River	Options limited by externally imposed waste form requirements, other policy requirements, and cost and schedule constraints
6. Interim storage at SRS	DOE-Savannah River	Currently limited by availability of storage space at Savannah River

7. Transportation to repository	DOE-Savannah River and DOE-Yucca Mountain	Will be limited by availability of shipping capacity at Savannah River and receiving capacity at DOE-Yucca Mountain
8. Storage at, and emplacement in, the repository	DOE-Yucca Mountain	Will be limited by availability of storage and emplacement capacity at the repository
9. D&D or reuse of treatment and storage facilities at SRS	DOE Office of Environmental Management, Environmental Restoration Program	Will depend on the nature of the treatment facilities and future needs at Savannah River

Notes: SRS = Savannah River site; RBOF = Receiving Basin for Offsite Fuels.

objectives of the phased strategy should be to maximize the probability of program success, minimize overall costs, and protect the program against the down-side risks from changes over which it has little or no control. The major programmatic decisions that must be made by DOE-Savannah River include the selection of one or more options for treating aluminum spent fuel and also the selection of a design for the treatment, storage, and shipping (TSS) facilities. The criteria for the decision-making process include the effectiveness of the treatment process, cost, schedule, compliance with applicable environmental health and safety standards, and consistency with other applicable policies. The options selected and facilities constructed also must be matched appropriately to the front (spent fuel generation) and back (disposal and D&D) ends of the overall disposal program (Table 5.1).

PHASED DECISION AND IMPLEMENTATION STRATEGY FOR TREATMENT OPTION SELECTION

DOE-Savannah River appears to recognize the importance of a phased decision-making strategy and is already applying it to individual parts of its program. However, a systems-oriented strategy is needed in the treatment program to ensure that technically sound and cost-effective decisions are made and implemented in a timely manner. Several examples of important considerations for a phased strategy of treatment option selection are summarized in Table 5.2 and discussed below. These examples are presented for illustrative purposes only and do not necessarily represent all of the significant considerations that apply to this program—although they are representative of the significant considerations.

Spent Fuel Generation

The quantity and type of aluminum spent fuel to be received and treated at Savannah River is one of the most significant factors in selecting a treatment option. Yet based on the information received during the course of this study, DOE-Savannah River does not appear to have a reliable estimate of the total inventory, particularly that part of the inventory in the "tail" of the receipt schedule shown in Figure 5.1. Nor does DOE appear to have considered the full range of scheduling options for returning this fuel to Savannah River for treatment. With the one exception noted in Chapter 4, DOE-Savannah River appears to have taken the return receipt schedule shown in Figure 5.1 as a given and has done relatively little thinking to date about how changes in this schedule could impact the treatment program and its cost.[1]

The aluminum spent fuel inventory can be divided into two components for the purposes of selecting a treatment option. The first component is the spent fuel that now exists at DOE-Savannah River or in offsite locations or that is likely to exist and be available for shipment to Savannah River by about 2015. This inventory would include all of the aluminum spent fuel now in storage at Savannah River, all of the foreign research reactor fuel,[2] the aluminum spent fuel now in storage at the Idaho Engineering and Environmental Laboratory (INEEL), and other domestic research reactor fuel that will be generated and available for shipment prior to about 2015 (Figure 5.1). This might be referred to as *legacy fuel*. The second component is the fuel that will be available for shipment to Savannah River after about 2015. According to Figure 5.1,

---

[1] Although it appears from the most recent program update (WSRC, 1998) that DOE-Savannah River is beginning to incorporate spent fuel receipt schedule planning in its treatment option selection decision.

[2] As shown in Figure 5.1, all of the foreign research reactor fuel is scheduled for shipment to Savannah River by 2009. No new shipments are expected to be added to this inventory.

TABLE 5.2 Objectives and Constraints in Aluminum Spent Fuel Program

Program Component	Objectives	Constraints	Major Uncertainties
Generation of spent fuel	Because treatment options and facility costs will be determined by size and characteristics of spent fuel inventory to be received at Savannah River, do not make major programmatic decisions that involve assumptions about the part of the inventory that is poorly constrained	*BUT* do consider options for treating that part of the inventory that is presently in storage at SRS or that will be received there in the next few years	The size of the post-2015 spent fuel inventory
Spent fuel receipt at SRS	Because the cost of time is so expensive, accelerate the return of aluminum spent fuel to SRS	*BUT* not so quickly that new receipt or storage facilities are required	Maximum receipt and storage capabilities at SRS
Spent fuel storage at SRS	Given that dry storage is less expensive and has fewer requirements for maintenance and servicing, do not rewet any fuel shipped dry to Savannah River	*BUT* review all dry fuel shipments for safety, and treat damaged and degraded fuel by conventional reprocessing	Condition of fuel shipped to SRS
Spent fuel storage at SRS	Given that dry storage is less expensive and has fewer requirements for maintenance and servicing, proceed to treat the fuel in wet storage as soon as possible	*BUT* make sure that treatment will produce an acceptable repository waste form so that retreatment is not required	Acceptability of some aluminum spent fuels for direct co-disposal

Spent fuel treatment	Arrange to conventionally reprocess as soon as possible all fuel that has a health- or safety-hazard label	*BUT* do not limit conventional reprocessing to today's hazard labels; include all fuel likely to pick up such a label within the project time horizon (e.g., if it will deteriorate within the operational life of the Canyons, reprocess it now)	Amount of fuel that will require conventional reprocessing for safety reasons
	Arrange to conventionally reprocess as soon as possible odd-sized fuel that will be difficult to treat by direct co-disposal	*BUT* do not treat this fuel at the expense of the hazard labels	Policy acceptability of conventional reprocessing
	Cost of time is so significant that it may make economic sense to treat all fuel now with a process that will ensure an acceptable waste form	*BUT* not if this approach does not save time and money in the long run	Acceptability of some aluminum spent fuels for direct co-disposal
	Treatment option selected should produce an acceptable waste form	*BUT* do not be restrained from considering treatment options that produces an untested waste form if an option is advantageous for other reasons	Acceptability of melt and dilute waste form

Table continued on next page

TABLE 5.2 Continued

Program Component	Objectives	Constraints	Major Uncertainties
D&D or reuse of treatment and storage facilities at Savannah River	Modify current facilities at SRS to handle, treat, and interim-store aluminum spent fuel	*BUT* not if this approach does not save time and money in the long run	D&D costs
Overall system	Consider the trade-off among process costs, container costs, storage costs, shipping costs, and repository costs	*BUT* recognize that the cost of time may void any other perceived cost benefits	Overall costs in light of the foregoing constraints

Note: SRS = Savannah River site.

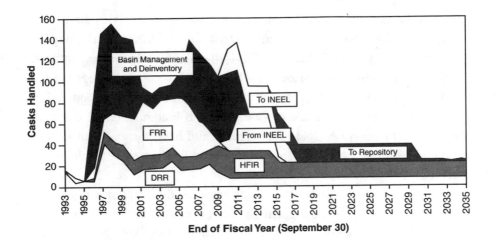

FIGURE 5.1 Projected receipts and transfers of aluminum spent fuel at Savannah River. NOTE: FRR = foreign research reactor fuel; DRR = domestic research reactor fuel; HFIR = High-Flux Isotope Reactor fuel; from INEEL = aluminum spent fuel now in storage at the Idaho National Engineering and Environmental Laboratory that will be shipped to Savannah River for treatment; to INEEL = non-aluminum spent fuel at Savannah River that will be shipped to INEEL for treatment; to repository = treated aluminum spent fuel that will be shipped to a repository for disposal; SRS = Savannah River Site. SOURCE: WSRC (1998), Figure 2.

this would include only HFIR[3] and domestic research reactor fuel and probably represents spent fuel that is not yet in existence. This might be referred to as *future fuel*.

The grouping of fuel into these two components is suggested because the inventory of aluminum spent fuel in the second category appears to be poorly defined.[4] Much of this fuel has not yet been fabricated or shipped to the reactors where it will be used. Changes in the need for research reactor capacity, changes in the number of operating reactors, or changes in future fuel designs and materials could decrease the amount of generated spent fuel. This could have a significant impact on the size and cost of treatment facilities and the treatment technologies selected.

Given that the first component comprises the great majority of the total inventory of aluminum spent fuel to be treated at Savannah River, any facilities constructed to treat this component will likely be more than adequate to handle the post-2015 inventory. Consequently, DOE-Savannah River should be able to tailor its decisions on treatment options and facilities to the legacy component of the aluminum spent fuel inventory, and at the same time it can obtain additional information to improve the estimates of the post-2015 inventory for a later treatment decision.

Spent Fuel Receipt and Storage

As part of the phased decision strategy on the treatment option for aluminum spent fuel, the fuel receipt and storage schedule will have to be considered, and one of the important programmatic factors in this

---

[3] High Flux Isotope Reactor, located at the DOE Oak Ridge site.

[4] The estimates of spent fuel receipt rates beyond about 2010 (Figure 5.1) are based on historical receipt rates and do not take into account future decisions concerning reactor operations, such as decisions to shut down reactors as they reach the ends of their operating lives.

schedule is the high *cost of time*.[5] The prompt shipment of all aluminum spent fuel to Savannah River for treatment might require the purchase of additional shipping casks but could significantly reduce overall costs, especially if the pre-2015 inventory could be returned to Savannah River in time for treatment by conventional reprocessing as discussed below. DOE-Savannah River has recognized the potential advantages of accelerating fuel returns and is beginning to consider such options in its program planning as evidenced by its most recent program update (WSRC, 1998). There is, however, an important trade-off between accelerated receipt schedules and the cost of handling and storage facilities at Savannah River, so shipment and treatment must be phased to minimize the need for new facilities. Additionally, given the high cost of operating the wet basins, they need to be emptied as quickly as possible. Decisions on receipt and storage rates will depend to a great extent on the treatment option(s) selected, as noted below.

Treatment and Interim Storage

There does not appear to be enough information at present to determine the acceptability of some of the options for treating aluminum spent fuel. As noted in Chapter 3, some aluminum spent fuel may not meet waste acceptance criteria for direct co-disposal in the repository. Also, for policy reasons, conventional reprocessing may be limited to aluminum spent fuel that represents health or safety hazards, and the reprocessing facilities at Savannah River may not be available after 2002.

---

[5] The cost of time can be thought of as the operational costs that are unrelated to actual production activities. These would include management and administrative costs, costs of supporting workers in a stand-by mode, and other operational costs that are time related rather than production or throughput related, for example, certain types of maintenance costs. To the first order, these operational costs are fixed per unit of time, consequently, cost is approximately proportional to time.

DOE-Savannah River must fill these information gaps before a treatment option can be selected.

Based on information received by the P.I. during the course of this study, there does not appear to be a technical basis for rejecting conventional reprocessing as an option for treating of aluminum spent fuel from foreign and domestic research reactors. Conventional reprocessing is a proven and reliable spent fuel treatment technology based on over 300 plant-years of operation worldwide, and the necessary treatment facilities (the F and H Canyons and the Defense Waste Processing Facility [DWPF]) are operating at Savannah River and are being used to treat aluminum spent fuel from research and production reactors.

The alternative cost study prepared by Westinghouse Savannah River Company (WSRC, 1997b) suggested that conventional reprocessing was a cost-effective treatment option when compared with direct co-disposal and melt-and-dilute treatment, the two primary treatment alternatives considered by the Task Team. However, the cost estimates for these three treatment alternatives have not been independently validated in this or any other study. Although it is difficult to make quantitative comparisons between a proven treatment technology such as conventional reprocessing and some of the other unproven treatment technologies considered by the Task Team, it is clear that the cost, performance, and safety of unproven technologies have much greater uncertainties than those of a demonstrated technology such as reprocessing. The common-basis cost and performance comparison of the two primary treatment alternatives (direct co-disposal and melt and dilute treatment) and conventional reprocessing, which was recommended in Chapter 2 of this report, will enable DOE-Savannah River to determine whether conventional reprocessing is an appropriate treatment option for this fuel.

The concern with conventional processing appears to be mainly one of policy and is related to the use of reprocessing for waste management generally rather than any specific concern about reprocessing this particular fuel type. Current U.S. nonproliferation policy does not encourage the civil use of plutonium. Accordingly, the United States "does not itself engage in plutonium reprocessing for either nuclear power or nuclear explosive purposes."[6] The P.I. notes that plutonium separation is not a significant problem with conventional reprocessing of enriched aluminum spent fuel from research reactors. There is less plutonium in this fuel in comparison to commercial spent fuel owing to its high ^{235}U enrichment, and separation of plutonium is not a required part of reprocessing treatment. The plutonium can be left in the liquid waste stream along with the fission products for later vitrification in glass. For aluminum spent fuel, the ^{235}U separated during conventional processing represents a potential proliferation hazard, but it can be diluted with ^{238}U within the reprocessing facility to make LEU. Moreover, Savannah River is a weapons material secure site and will remain so for the duration of this program.

The reprocessing of aluminum spent fuel also does not appear to be in conflict with the DOE decision to phase out reprocessing at Savannah River (DOE, 1992a). The Highly Enriched Uranium Task Force noted in its predecisional draft report (DOE, 1992b; see Chapter 2) that the need for reprocessing for long-term DOE spent fuel management was unclear at present and that DOE should evaluate the near-term operational requirements to bring its facilities to a condition for transfer to the Office of Environmental Management for potential future operations. Indeed, as noted elsewhere in this report, DOE has or plans to reprocess some of its aluminum spent fuel in the Canyons at Savannah River because of safety concerns.

---

[6] The quote is taken from the White House Fact Sheet entitled *Nonproliferation and Export Control Policy* dated September 27, 1993. The fact sheet is based on Presidential Decision Directive 13, which is classified and was not reviewed in this study.

Additionally, the final environmental impact statement (EIS) for foreign research reactor fuel notes that the use of conventional reprocessing for treatment of foreign research reactor fuel would not be inconsistent with current U.S. policy (DOE, 1996c, p. 24):

> *DOE is aware that the inclusion of chemical separation* [conventional reprocessing] *within the preferred alternative could be interpreted by some nations, organizations and persons as a signal of endorsement of the use of reprocessing as a routine method of waste management for spent nuclear fuel or a reversal of U.S. policy on reprocessing. This would not be an accurate interpretation. The U.S. policy regarding reprocessing was established in Presidential Decision Directive 13. DOE and the Department of State have determined that this preferred alternative is not inconsistent with that policy. . . . The independent study[7]* [to be undertaken by DOE] *will review the policy, technology, cost and schedule implications for reprocessing foreign research reactor spent nuclear fuel to determine whether reprocessing of foreign research reactor spent nuclear fuel is justified for other than health and safety reasons.*

---

[7] As noted in Chapter 2, the foreign research reactor EIS (DOE, 1996c) called for DOE to commission or conduct an independent study of the nonproliferation and other implications of reprocessing spent nuclear fuel from foreign research reactors. Based on the description of this ongoing study, which was provided by Jon Wolfstal of DOE's Office of Arms Control and Nonproliferation at the second information-gathering session, it is an in-house study rather than a commissioned independent study. See additional comments by consultant David Rossin in Appendix D.

The acceptability of conventional processing might be increased if it were redesigned as a reprocess-and-dilute operation, in which spent fuel is conventionally reprocessed and the separated ^{235}U is diluted with ^{238}U to produce low enriched uranium (LEU) before it leaves reprocessing facility.[8] In the event that conventional processing is not found to be acceptable, it may still be possible to processes the degraded and damaged fuel (i.e., health and safety hazard "labels") as it is received at Savannah River and perhaps the odd fuel sizes that would be expensive to treat by other methods. As noted in Table 5.2, the definition of labeled fuel should be expanded to include damaged and degraded fuel that will be received in future shipments to Savannah River and fuel in danger of deteriorating while in storage at Savannah River.

If conventional reprocessing is not acceptable, then another treatment option will have to be used to treat the pre-2015 inventory of aluminum spent fuel. The selection of a treatment option should be based on a complete systems lifetime cost comparison of those options that meet regulatory requirements. According to the system cost estimates provided in the alternative cost study (WSRC, 1997b; see Table 4.3 in this report), either treatment option will entail about the same costs (the higher costs of melt and dilute treatment are offset by the higher disposal costs for direct co-disposal). Thus, the treatment option selected will depend to a great extent on the acceptability of the waste form for disposal in the repository. All else being equal, the high cost of time may favor the selection of melt and dilute treatment because the acceptability of a direct co-disposal waste form for some aluminum spent fuel is potentially problematical (Chapter 3). Melt and dilute treatment will entail more handling and processing of the fuel, but the waste form characteristics can be altered to make it suitable for repository disposal and, additionally, fewer waste containers will be produced.

The cost of a TSS facility to support any of the treatment options may be reduced significantly by phasing receipt and treatment of spent

---

[8] This dilution could in fact be done at almost any step of the process.

fuel or by utilizing existing facilities. For example, if conventional processing treatment is selected, it might be possible to process the pre-2015 inventory by utilizing one or both of the existing wet basins for all storage. If additional storage space is needed, it may be more cost-effective to modify an existing facility at the Savannah River site as an off-loading facility and to coordinate fuel receipt and processing schedules so that additional new storage is not required and double handling[9] of the fuel is minimized.

Similarly, it may be possible to modify existing facilities for melt and dilute or direct co-disposal treatment. For direct co-disposal treatment, it also might be possible to reduce the size of the TSS facility by drying and packaging, at INEEL, the aluminum spent fuel now in storage at that site—perhaps in the facility that will be built to process the zircaloy fuel now stored there. It does not appear to be optimum from the standpoint of either safety[10] or cost to ship the fuel from INEEL to Savannah River for drying and packaging and then to ship the packaged fuel from Savannah River to the repository for disposal. Some renegotiation of current agreements with the State of Idaho may be required to pursue this treatment option.

DOE-Savannah River recognizes that a repository may not be available to receive the treated fuel well into the twenty-first century and is prudently planning for several decades of onsite interim storage. Dry interim storage has lower operating costs and requires less maintenance and servicing than wet storage, the current storage medium for unprocessed spent fuel. Thus, it would be advantageous to treat the spent fuel as soon as possible and to put the road-ready canisters into interim storage until they can be shipped to the repository. Of course, the smaller the number of treated canisters, the less expensive it will be to construct

---

[9] For example, moving the fuel from the receiving facility into wet storage, then removing the fuel from wet storage at a later date to treat it.

[10] The risk of a conventional traffic accident during fuel shipping operations will probably dominate the transportation risk.

and maintain an interim storage facility. As noted above, it might even be possible to modify existing facilities, such as a drained wet basin, for such storage.

Transport and Repository Disposal

There appears to be little flexibility on decision-making in the transport and repository disposal parts of the system. The shipping schedule for treated aluminum spent fuel will be determined by the repository program, and the treatment option selected will determine the number of canisters to be shipped. Transportation and disposal costs will be lowest for those treatment options that produce fewer numbers of canisters. Of course, total systems costs will be reduced by shipping the fuel to the repository as soon as practical.

Decontamination and Decommissioning

Even though the D&D costs of facilities built to receive, handle, treat, and store aluminum spent fuel at Savannah River will not be charged directly to the spent fuel treatment component of the program, the costs will be borne ultimately by U.S. taxpayers. Thus, the decision strategy should consider the costs of D&D as part of the total life-cycle costs of the treatment program. The use of current facilities to treat the aluminum spent fuel (e.g., use of the Canyons and the DWPF for conventional reprocessing) or the modification of existing facilities will likely entail the lowest D&D costs. New facilities will entail additional D&D costs.

Post-2015 Aluminum Spent Fuel Inventory

There does not appear to be any reason at this time to make a decision about the disposition of the post-2015 inventory of aluminum spent fuel. That inventory could well be different in size (most likely smaller) and composition than currently anticipated, therefore, treatment options that do not appear to be available today may in fact be available when the time comes to treat this fuel. In addition, facilities might become available elsewhere in the DOE complex to deal with this small residual inventory. As suggested previously, additional efforts should be made to obtain better estimates of the size and characteristics of this waste stream before a treatment decision is made.

PATH FORWARD

DOE-Savannah River is doing a commendable job of collecting data for decision making on many of the individual components of its treatment option selection program. In addition, DOE-Savannah River is in the process of defining a decision strategy for selecting and implementing a treatment option for aluminum spent fuel. The decision strategy is in the early stages of formulation, and the P.I. did not ask for or receive detailed briefings on this strategy during the course of the study.

As part of this decision strategy, it is recommended that DOE-Savannah River conduct a complete systems review to identify and understand the relationships among the various components of the aluminum spent fuel disposal program shown in Table 5.1. DOE-Savannah River also is encouraged to apply a phased strategy for selecting and implementing a treatment option for aluminum spent fuel that takes into account the considerations given in Table 5.2 and discussed above. This phased approach will support the analysis required in the environmental impact statement and will lead to a more credible EIS-ROD (record of decision) and a more successful and cost-effective path forward for the aluminum spent fuel treatment program.

REFERENCES

Argonne National Laboratory (ANL). 1963. Description and Proposed Operations of Fuel Cycle Facility for the Second Breeder Reactor (EBR-II). ANL-6605. Chicago, Illinois: Argonne National Laboratory.

Civilian Radioactive Waste Management System (CRWMS). 1997a. Total System Performance Assessment Sensitivity Studies of U.S. Department of Energy Spent Nuclear Fuel. A00000000-01717-5705-00017, Rev. 01. September 30, 1997. Las Vegas, Nevada: TRW.

Civilian Radioactive Waste Management System (CRWMS). 1997b. Evaluation of Codisposal Viability for Aluminum-Clad DOE-Owned Spent Fuel: Phase 1. Intact Codisposal Canister. BBA000000-01717-5705-00011, Rev. 01. August 15, 1997. Las Vegas, Nevada: Civilian Radioactive Waste Management System

Howell, J. P. 1997. Fission Product Release From Spent Nuclear Fuel During Melting (U). WSRC-TR-97-0112 (U). Aiken, South Carolina: Westinghouse Savannah River Company.

Jonke, A. A. 1965. Process studies on the recovery of uranium from highly enriched uranium alloy fuels. Argonne National Laboratory Chemical Engineering Division Semiannual Report, January-June 1965. ANL-7055. Chicago, Illinois: Argonne National Laboratory.

Lam, P. S., R. L. Sindelar, and H. B. Peacock, Jr. 1997. Vapor Corrosion of Aluminum Cladding Alloys and Aluminum-Uranium Fuel Materials in Storage Environments (U). WSRC-TR-97-0120. Aiken, South Carolina: Westinghouse Savannah River Company.

Large, W. S. and R. L. Sindelar. 1997. Review of Drying Methods for Spent Nuclear Fuel (U). WSRC-TR-97-0075 (U). Aiken, South Carolina: Westinghouse Savannah River Company.

National Research Council (NRC). 1995. An Assessment of Continued R&D into an Electrometallurgical Approach for Treating DOE Spent Nuclear Fuel. Washington, D.C.: National Academy Press.

Research Reactor Spent Nuclear Fuel Task Team (Task Team). 1996. Technical Strategy for the Treatment, Packaging, and Disposal of Aluminum-Based Spent Nuclear Fuel. Prepared for U.S. Department of Energy Office of Spent Fuel Management (2 volumes).

Sindelar, R. L., H. B. Peacock, Jr., P. S. Lam, N. C. Iyer, and M. R. Louthan, Jr. 1996. Acceptance Criteria for Interim Dry Storage of Aluminum-Alloy Clad Spent Nuclear Fuels (U). WSRC-TR-95-0347 (U). Aiken, South Carolina: Westinghouse Savannah River Company.

TRW Environmental Safety Systems, Inc. (TRW). 1997a. Mined Geologic Disposal System Waste Acceptance Criteria. B00000000-01717-4600-00095, Rev 00. September 1997. Las Vegas, Nevada: TRW Environmental Safety Systems Inc.

TRW Environmental Safety Systems, Inc. 1997b. (TRW). OCRWM Data Needs for DOE Spent Nuclear Fuel. DI: A00000000-01717-2200-00090, Rev. 02. September, 1997. Vienna, Virginia: TRW Environmental Safety Systems, Inc.

U.S. Department of Energy (DOE). 1992a. Memorandum from DOE Secretary James D. Watkins entitled, "Highly Enriched Uranium Task Force Report." February 24, 1992. Washington, D.C.: U.S. Department of Energy.

U.S. Department of Energy (DOE). 1992b. Task Force Study on DOE Spent Fuel Reprocessing, Summary Report (Predecisional Draft). February 1992. Washington, D.C.: U.S. Department of Energy.

U.S. Department of Energy (DOE). 1995. Programmatic Spent Fuel Management and Idaho National Engineering Laboratory Environmental Restoration and Waste Management Programs Final

Environmental Impact Statement. Idaho National Engineering Laboratory. DOE/EIS-0203-F (14 volumes). Idaho Falls, Idaho: U.S. Department of Energy.

U.S. Department of Energy (DOE). 1996a. Integrated Data Base Report— 1995: U.S. Spent Nuclear Fuel and Radioactive Waste Inventories, Projections, and Characteristics. DOE/RW-0006, Rev. 12. Washington, D.C.: Office of Environmental Management.

U.S. Department of Energy (DOE). 1996b. DOE-Owned Spent Nuclear Fuel Technology Integration Plan. DOE/SNF/PP-002. May 1996. Idaho Falls, Idaho: U.S. Department of Energy.

U.S. Department of Energy (DOE). 1996c. Final Environmental Impact Statement: Proposed Nuclear Weapons Nonproliferation Policy Concerning Foreign Research Reactor Spent Nuclear Fuel. Summary. DOE/EIS-0218F. February 1996. Washington, D.C.: U.S. Department of Energy.

U.S. Nuclear Regulatory Commission (USNRC). 1997. Code of Federal Regulations, Title 10, Part 60, Disposal of High-Level Radioactive Wastes in Geologic Repositories. Washington, D.C.: U.S. Government Printing Office.

Westinghouse Savannah River Company (WSRC). 1997a. Alternative Aluminum Spent Nuclear Fuel Treatment Technology Development Status Report (U). WSRC-TR-97-00345(U). October 1997. Aiken, South Carolina: Westinghouse Savannah River Company.

Westinghouse Savannah River Company (WSRC). 1997b. Savannah River Site Aluminum-Clad Spent Nuclear Fuel Alternative Cost Study (U). WSRC-RP-97-299 Rev. 1. December 1997. Aiken, South Carolina: Westinghouse Savannah River Company.

Westinghouse Savannah River Company (WSRC). 1997. Savannah River Site FY 1998 Spent Nuclear Fuel Interim Management Plan. WSRC-RP-97-00922. November 1997. Aiken, South Carolina: Westinghouse Savannah River Company.

APPENDIX A

Study Request Letter from DOE-Savannah River

Department of Energy
Savannah River Operations Office
P.O. Box A
Aiken, South Carolina 29802
JUN 1 9 1997

Dr. Michael Kavanaugh, Chair
Board on Radioactive Waste Management
National Research Council
2101 Constitution Ave., NW
Washington, DC 20418

Dear Dr. Kavanaugh:

SUBJECT: Review of Aluminum-Based Spent Nuclear Fuel Disposition Technology

The Department of Energy (DOE) is currently performing research and development work as well as cost and timing evaluations to aid in making decisions regarding the selection of a technology that could be utilized to prepare aluminum-based spent nuclear fuel for placement in a geologic repository. This fuel is constructed of aluminum (either as cladding or as a uranium-aluminum alloy fuel material) and is therefore more vulnerable to corrosion than commercial spent nuclear fuel. This aluminum-based spent nuclear fuel is from research and test reactors in the United States, or from foreign research reactors. Currently, Savannah River Site (SRS) has approximately 17 cubic meters aluminum-based spent nuclear fuel stored in wet basins. This quantity is expected to increase to 255 cubic meters by 2035.

DOE is currently studying two different disposition technologies (direct co-disposal, and melt and dilute) for interim storage at SRS, as well as preparation of the fuel for shipment to the repository. The choice of these two alternatives was based on the recommendation of the Research Reactor Spent Nuclear Fuel Task Team. The team evaluated and compared eleven different treatment technologies that could be utilized in place of conventional processing. The team recommended that DOE proceed with the parallel development of two treatment technologies. Since the team issued its report (reference: Technical Strategy for the Treatment, Packaging and Disposal of Aluminum-Based Spent Nuclear Fuel, June 1996), DOE has been studying the advantages and disadvantages of these two technologies to support a December 1997 technology decision.

The National Research Council has a proven track record of providing the DOE with scientific analysis of complex technical issues; notably, the recent report of alternatives for the removal and disposition of molten salt reactor and experiment fluoride salts. Therefore, the Department requests that the National Research Council evaluate our on-going work on aluminum-based spent nuclear fuel disposition to provide recommendations on improvements in our process.

Dr. Kavanaugh   2   **JUN 1 9 1997**

To be most useful to the department, the council's evaluations should include:

- A review of the alternative technologies considered by the Department for the disposition of aluminum-based spent nuclear fuel, with a focus on non-chemical-based separation technologies, and the identification of other technologies that should be considered.

- The suitability of the technologies that are currently the focus of the Department's research efforts.

- The Department's work to identify and address likely geologic repository waste acceptance criteria that would be applicable to the Department's aluminum-based spent nuclear fuel.

- The cost and timing aspects of implementing a likely spent nuclear fuel disposition technology.

I request to see the council's recommendations by December 15, 1997. Mr. Karl Waltzer of my staff will be the principal point of contact for coordinating your investigation. We will be happy to provide briefings or any available documentation the council may request.

Please contact Karl Waltzer of my staff at extension (803) 952-4121 if you or your staff have any questions.

Sincerely,

A. Lee Watkins
Assistant Manager for Material
and Facility Stabilization

TECH:KEW:mt

UD-97-0109

cc: Dr. Kevin D. Crowley

APPENDIX B

Meeting Agendas and Participants

FIRST MEETING
November 4-5, 1997
Aiken, South Carolina

November 4, 1997 (Tuesday)

Open Session

8:30 a.m.	Introduction of meeting participants	Milton Levenson Kevin Crowley
8:45 a.m.	Review of National Research Council project procedures	Kevin Crowley
9:15 a.m.	Tentative project schedule	Milton Levenson
9:45 a.m.	Overview of technical strategy for treatment, packaging, and disposal of aluminum-based spent fuel	DOE-Savannah River
10:00 a.m.	Break	
10:15 a.m.	Overview of cost and schedule for treatment, packaging, and disposal of aluminum-based spent fuel	DOE-Savannah River
10:45 a.m.	Licensing and waste acceptance	DOE-Savannah River
11:45 a.m.	Discussion and questions	

119

12:15 p.m.	Lunch	
1:30 p.m.	Review of technical options for treatment, packaging, and disposal of aluminum-based spent fuel	DOE-Savannah River
3:15 p.m.	Break	
3:30 p.m.	Review of technical options for treatment, packaging, and disposal of aluminum-based spent fuel (continued)	
4:30 p.m.	Additional information needed and plans for the December meeting	Milton Levenson Kevin Crowley
5:00 p.m.	Adjourn	

Wednesday, November 5 (Wednesday)

| 9:00 a.m. | Tour of Savannah River facilities | Milton Levenson Kevin Crowley |

SECOND MEETING
December 2-3, 1997
Augusta, Georgia

December 2, 1997 (Tuesday)

Open Session

8:00 a.m.	Welcome and introduction of participants; plan and objectives of meeting	Milton Levenson Kevin Crowley
8:20 a.m.	Overview of DOE's aluminum-based spent fuel program	Karl Waltzer, DOE
8:50 a.m	Overview of research reactor task team study and results	Jack DeVine, Polestar
9:50 a.m.	Alternate technology program	Mark Barlow, WSRC
10:15 a.m.	Break	
10:30 a.m.	Review of waste form, waste container, and waste acceptance criteria and their impact on selecting a disposition option for implementation	Hugh Benton, TRW
11:30 a.m.	Review of handling and processing issues and their impact on selecting a disposition option for implementation	Dick Murphy, WSRC
12:15 p.m.	Lunch	

1:30 p.m.	Review of metallurgy and corrosion issues and their impact on selecting a disposition option for implementation	Natraj Iyer, WSRC
2:15 p.m.	Review of criticality issue and its impact on selecting a disposition option for implementation	Peter Gottlieb, TRW
3:00 p.m.	Review of proliferation issue and its impact on selecting a disposition option for implementation	Jon Wolfstal, DOE
3:45 p.m.	Break	
4:00 p.m.	Costs and schedules for implementation of the comparative case and the 11 other disposition options	Joe Krupa, WSRC
4:45 p.m.	Review of additional work to be completed on the two preferred and one backup options, and path forward for selecting a disposition option for implementation	Natraj Iyer, WSRC
5:30 p.m.	Adjourn	

December 3, 1997 (Wednesday)

Open Session

8:30 a.m.	Small group interactions among consultants, Savannah River staff, and other interested participants to discuss remaining issues from yesterday's presentations.
12:00 noon	Lunch

1:15 p.m.   Reconvene in plenary session (consultants
                  will be asked to discuss important observations
                  and needs, if any, for additional information)

3:00 p.m.   Adjourn

Consultants Who Participated In The Second Meeting

Francis Alcorn, BWX Technologies, Lynchburg, Virginia
Maurice Angvall, Bechtel, Chico, California (retired)
Robert Bernero, U.S. Nuclear Regulatory Commission, Gaithersburg,
Maryland (retired)
Joseph S. Byrd, University of South Carolina, Lexington (retired)
Robert L. Dillon, Hanford, Richland, Washington (retired)
Brian Estes, NAVFACENGCOM & Westinghouse Hanford,
Williamsburg, Virginia (retired)
Harry Harmon, NUKEM Nuclear Technologies, Columbia, South
Carolina
Valerie Putman, INEEL, Lockheed Martin Idaho Technologies, Idaho
Falls
David Rossin, Rossin and Associates, Los Altos, California
Paul G. Shewmon, Ohio State University, Columbus (retired)
Richard I. Smith, PNNL, Kennewick, Washington (retired)

APPENDIX C

Charge to Consultants

Note: The following material was sent to the consultants who attended the second information-gathering meeting (see Appendix B).

We are pleased that you have agreed to participate as a technical expert in the National Research Council (NRC) project entitled *Technical Options for Disposition of Aluminum-Based Spent Nuclear Fuel,* which is being undertaken by the NRC's Board on Radioactive Waste Management for the U.S. Department of Energy's (DOE's) Savannah River Office. This document contains a brief set of instructions to guide your preparation for and participation at our meeting in Augusta, Georgia, on December 2-3, 1997. We will discuss these instructions in more detail at the beginning of the December meeting, but you should feel free to call Kevin Crowley (202-334-3066) if you have any immediate questions or concerns.

The objective of the December meeting is to obtain the information needed to develop a National Research Council report that fully addresses the statement of task for this project (the task statement is included in this package). To accomplish this objective, we have invited about a dozen experts to the meeting to provide advice on the technical issues in the task statement. Given the time constraints for this meeting and the overall project (a final report will be issued in March 1998), we have developed the following set of questions to focus the presentations and discussions. You will be assigned a subset of these questions and will be asked to gather the necessary information at the meeting to answer them. Time will be set aside at the meeting for presentations by DOE on the technical issues and for small group discussions so that you can gather the information you need to fully address your questions. We will ask each of you to provide us with a 5-10 page write-up of your answers and other relevant comments before the end of December. We plan to include these write-ups in the appendix of the final NRC report and to credit you in that report for your participation as a technical expert.

Please keep the following two thoughts in mind as you review and answer these questions. First, we ask that you focus your efforts on those aspects of DOE's program that are under review in this project. DOE's aluminum-based spent fuel program includes a variety of activities, including the ingathering of fuel from foreign research reactors, the shipping of domestic research reactor fuel to Savannah River, storage of the foreign and domestic research reactor fuel at Savannah River, preparation or processing to ready the fuel for the repository, loading the processed fuel in a "road-ready" canister for shipment to the repository, interim storage of the processed fuel until the repository opens, shipment to the repository, and emplacement in the repository. This study will not review all of these activities. As explained in the statement of task, the objective of this project is to review *only* the processing or preparation options, the resulting waste form properties, the canister as it might be affected by the waste form (the canister qualification for repository conditions is outside the scope of this review), and interim storage plus any incremental effects (e.g., criticality effects) that may occur in the repository due to the addition of the waste form. Second, in addressing the questions you should clearly differentiate between fact and informed opinion and provide references to previous work wherever possible to back up your facts and opinions. We do of course want to have your informed opinions as well as the facts, but we need a clear differentiation to develop an accurate and balanced final report that will pass muster in the very rigorous NRC review process.

The list of questions and expert assignments is given below. Please review these as soon as possible and let us know whether we have assigned you the right subset of questions and whether there are other questions we should add to this list.

Thank you.

Questions To Experts

Criticality

1. What are the significant criticality issues that must be considered during processing, interim storage of the waste form after processing, and shipment of the waste form to a repository? Has DOE adequately addressed these issues in its technology planning?

2. Do any of the waste forms produced by the alternative processing options pose significant internal or external criticality hazards in a repository—from material degradation either in the waste container or in the near field of the repository after the container is breached—*relative to commercial spent fuel or vitrified high-level waste?* Note: comments on the use of poisons or isotopic dilution are appropriate as are comments on filling the void space in the canister so as to limit the volume of water that could be present in case of canister leakage.

Proliferation

1. Has DOE considered the proliferation-diversion issue in its technology selection planning?

2. Will the waste forms resulting from any of the alternative processing options represent a more attractive target for proliferation diversion than spent commercial nuclear fuel? If so, how might such attractiveness be reduced or eliminated?

3. Are there any significant differences among the different waste forms with respect to their potential attractiveness for proliferation diversion?

Cost and Schedule

1. Are the cost data provided by DOE reasonably complete and transparent?

2. Are the cost and schedule estimates developed by DOE for the alternative processing options suitable as a basis for comparison and selection of one or more preferred alternatives?

3. Are the cost and schedule estimates developed by DOE for the alternative processing options suitable for budget planning purposes?

4. Has DOE considered the costs of program delays in its budget development or budget planning for this program?

5. Are the cost and schedule estimates for implementing the alternative processing options consistent with DOE procedures and systems? If not, has DOE identified what changes must be made to achieve its cost and schedule targets?

6. Are the cost and schedule milestones that are laid out in the Research Reactor Task Force Report for selecting and implementing an alternative processing option being met?

Corrosion and Metallurgy

1. Are DOE's plans for fuel handling, drying, and interim storage technically credible? Are these process steps adequate to prevent significant fuel corrosion?

2. For each of the processing options evaluated by DOE, are the processing steps used as the basis for assessment and comparison (other than direct co-disposal) technically credible? That is, are they likely to work as described and to produce the products and results assumed?

3. Since the amount of water in the repository is likely to be somewhat limited, would filling the canister void space with aluminum or some other sacrificial material make any difference in long-term corrosion of the waste form?

4. Will any of the waste forms resulting from any of the alternative processing options be likely to increase internal corrosion of a standard repository container compared to spent commercial fuel or vitrified glass logs? That is, are there likely to be interactions between the waste form and the inner container in excess of what would be expected for spent commercial fuel or vitrified glass logs?

5. What is the status of R&D activities at Savannah River on the melt and dilute and co-disposal options? Are the R&D activities appropriately focused and are they likely lead to useful outcomes?

Processing and Remote Handling

1. For each of the processing options evaluated by DOE, are the processing steps used as the basis for assessment and comparison (other than direct co-disposal) technically credible? That is, are they likely to work as described and to produce the products and results assumed?
2. Are there other processing options that should be considered by DOE for disposition of aluminum-based spent fuel?
3. Do the inner container designs appear adequate to contain the waste forms resulting from the various processing options? Are they overdesigned for the intended application?
4. Are DOE's basic material handling plans, pool use, and other facility needs reasonable, and are remote handling technologies available to meet these needs?
5. Are the technical requirements of the various alternative processing options sufficiently well defined so that reasonable judgments can be made about the likelihood of success of implementing them?
6. Are there large differences in likelihood of success of implementing the various processing options?

Regulatory Waste Acceptance

1. Has DOE-Savannah River identified the appropriate criteria for aluminum-based spent fuel from the draft waste acceptance criteria document that has been prepared by the Yucca Mountain program?
2. Which of the waste forms is likely to be most acceptable for disposal in a repository relative to commercial spent fuel and vitrified logs? For those waste forms that are unlikely to be acceptable, has DOE considered alternate processing options?
3. Are the waste acceptance criteria that have been identified by DOE suitable for selecting among the alternative processing options?
4. Is DOE-Savannah River making an adequate effort to stay current with changes in waste acceptance criteria?

APPENDIX D

Consultant Reports

Note: The principal investigator (P.I.) consulted with several expert consultants in the course of this study. At the request of the P.I., the following reports were contributed by 11 consultants who were invited to attend the second information-gathering meeting and two consultants who did not attend that meeting (see Appendixes B and C). Biographical sketches of the consultants are included in Appendix E.

The consultants who attended the second information gathering meeting had access to many (but not all) of the reports cited in Appendix F. Most significantly, none of the consultants had access to the predecisional draft of the Highly Enriched Uranium Task Force report (DOE, 1992b) discussed in Chapter 5.

The opinions, findings, conclusions, and recommendations provided in these reports represent the views of the consultants and do not necessarily represent the views of the P.I. or the National Research Council.

Topic: Proliferation Aspects of the Treatment Options
Consultant: Harold Agnew

Once the material is under U.S. or DOE custody under today's management, I see no proliferation risk from any of these spent or reprocessed fuel forms. Unless the enriched material is needed it will be a waste of money and resources to reprocess it under any proliferation scenario. Reprocessing will also add to our present waste disposal problems. Proper containment should be the only objective in any reprocessing or repackaging. For the longer term, if it were to be placed in the center of an overpack surrounded by five canisters of high-level vitrified waste and buried in the repository among thousands of overpacks of spent commercial reactor fuel and other canisters of vitrified waste I would not consider it a credible proliferation target.

As an aside, if one worries about proliferation using enriched uranium, one should be concerned about Deputy Aleksadr Belosokov's statement reported in the 12/12/97 (p. A3) *New York Times* that Russia will scrap the contract to sell processed enriched uranium from weapons to the United States Enrichment Corporation and make the material available worldwide.

With the USSR/Russia starting to renege on its sale of 500 metric tons of HEU^{235} (approximately ten times the amount of research reactor fuel) the question of final form for spent aluminum clad or alloyed research reactor fuel is moot. In my opinion, none of the proposed "recycled" forms are less or more susceptible to proliferation. In fact, I believe the less the fuel is "massaged" the better. Reprocessing will result in "muffs." It will be easier to account for the material if it is stored in its original form. It will save reprocessing costs, and accounting costs and produce no wastes, so it will be environmentally better. The real worry will be if Yeltsin leaves: commerce between Russia and Iran, Iraq, Pakistan, and others will be the real concern. HEU from Russia's reserves and from stockpile reductions are enormous. The issue of spent aluminum-clad or alloyed research reactor fuel is not worth considering with regard to changing its physical form. Leave it alone and account for it. The more you handle material, the greater the chance for mischief.

Topic: Proliferation Aspects of the Treatment Options
Consultant: John Ahearne

The principal proliferation concern relating to the aluminum-based fuels is that many are highly enriched uranium (HEU). For example, according to a DOE document, one shipment has over 100 kg of 93% enrichment (*Savannah River Site FY97 Spent Nuclear Fuel Interim Management Plan*, WSRC-RP-96-530, 21 October 1996, p. C-2). The significance is that "typical weapons-grade uranium is more than 90 percent U-235" (*Management and Disposition of Excess Weapons Plutonium*, National Academy Press, 1994, p. 30). Although recent concerns have focused on plutonium, HEU may be of greater concern because "plutonium can only be used in implosion weapons." However, "[h]ighly enriched uranium (in weapons, typically 90 percent U-235 or more) can be used in either gun-type nuclear weapons designs like that used at Hiroshima or in the more efficient implosion design" (*Management and Disposition of Excess Weapons Plutonium: Reactor-Related Options*, National Academy Press, 1995, p. 43). Thus, the spent fuel from many of the reactors, with HEU still a large part of the fuel, is a serious proliferation risk—IF taken by a group that would be able to extract the uranium from the highly radioactive fission products.

The protection provided by this radioactivity is the basis of what the National Academy has called "the spent fuel standard" (ibid., 1994, op. cit., p. 12). So long as the disposal option keeps the fission products with the HEU, this protective barrier remains. However, several of the options appear to include separation of the uranium, at least at one or another stage in the process. For example, in the electrometallurgical treatment, which some have recommended be retained as "a secondary and diverse backup" (*Technical Strategy for the Treatment, Packaging, and Disposal of Aluminum-Based Spent Nuclear Fuel: A Report of the Research Reactor Spent Nuclear Fuel Task Team*, Vol. 1, June 1996, p.78), the uranium is separated out (see flow diagram on p. 42, ibid., and several recent NRC reports focused on this process). Although the plan here apparently is to mix in depleted uranium to blenddown the HEU to LEU, the process does permit separation of nearly pure weapons-grade uranium. This would at a minimum require substantially tighter safeguards than other processes under consideration. Unless one of the options is chosen that blends down the HEU with depleted uranium, preferably without separating the HEU at any stage of the process, the issues surrounding disposal are essentially the same as those for the weapons-grade plutonium treated in the referenced NAS reports. This includes recognition that some of the more intense radioactive materials relied upon for self-protection have short half-lives (decades) so that if the spent fuel is in storage for many decades, the self-protection is weakened, increasing the need for safeguards.

Compared with commercial spent fuel, this fuel presents greater proliferation risks because of the HEU. If the fission product load is sufficient to match the spent-fuel standard for commercial fuel, then the risk for this research fuel would be basically the same as for the commercial fuel. (The proliferation concern with commercial fuel is the plutonium produced during power generation.) The safeguards proposed for commercial fuel would be necessary if the radioactive protective barrier is maintained for the aluminum-based research fuel. Diluting the HEU with depleted uranium would reduce the proliferation hazard and, depending on how the dilution was accomplished (i.e., actual mixing would be required), could reduce the safeguards required.

Topic: Nuclear Criticality Safety
Consultant: Francis M. Alcorn

This report is tendered by Francis M. Alcorn based on the following:

1. Participation in a technical meeting held in Augusta, Georgia on December 2 and 3, 1997. On Tuesday, December 2 there were ten technical presentations on various aspects of the project. Only one of these dealt directly with criticality issues. That presentation, by Peter Gottlieb of TRW, addressed phases 1 and 2 (out of 3) for a codisposal waste package in a repository; Peter's work is sponsored by the Office of Civilian Radioactive Waste Management (OCRWM). On Wednesday there was additional dialog with Peter.

2. Review of approximately 14 documents. Of these, four sets dealt with criticality issues:

a. Volumes I and II of "Technical Strategy for the Treatment, Packaging and Disposal of Aluminum-Based Nuclear Fuel" (June 1996).

b. "Alternative Aluminum Spent Nuclear Fuel Treatment Technology Development Status Report (U)," WSRC-TR-97-00345(U) (October 1997).

c. Six documents by OCRWM with the same primary report number of WBS: 1.2.2/QA.L; one of the reports was dated August 1997, while the other five were dated September 1997.

d. "Technical Strategy for the Management of INEEL Spent Nuclear Fuel" (March 1997) was reviewed; although it discussed criticality issues, it was of marginal value to this review.

3. Instructions from the Principal Investigator. These instructions were:

a. Focus only upon the processing or preparation options, the resulting waste form properties, the canister as it might be affected by the waste form (canister qualification for the repository is outside of this review), and interim storage plus any incremental effects that might occur in the repository due to the addition of the waste form.

b. Respond to two specific questions:

i. What are the significant criticality issues that must be considered during processing, interim storage after processing, and shipment to a repository? Has DOE adequately addressed these issues?

ii. Do any of the waste forms produced by the alternatives pose significant internal or external criticality hazards in a repository?

My general observation, based upon Peter Gottlieb's presentations and review of the six OCRWM reports, is that the Operating Contractors for OCRWM are performing very detailed and thorough criticality safety evaluations. The Savannah River status report mentioned above (2.b) is, from a criticality perspective, primarily a status report of OCRWM activity to date (section 4.5). Although one cannot argue with the technical quality of this work, the following should be noted:

1. This work primarily addresses canister performance in a repository (which is somewhat outside of the requested focus) and satisfying 10 CFR 60;

2. the work claims to have completed only two of three phases for the canister if the codisposal waste option is assumed; and

3. it is obvious that some of this work must be repeated since both the Peter Gottlieb presentation and the Savannah River status report talk of investigations in progress to select an appropriate neutron poison material for the codisposal canister.

It also appeared that the OCRWM contractors have performed a significant amount of criticality evaluations, while DOE/Savannah River staff has done very little in criticality evaluations to support processing, interim storage, and shipping for each of the alternatives. For the scope of this review it would appear, in my opinion, that Savannah River staff should have made the presentation on criticality. In my opinion the Savannah River staff is technically capable to complete their part of the project. Because so little had been done, criticality wise, for that part of the project under review it was difficult to do an appropriate review.

In response to the first specific question: Has DOE adequately addressed the significant criticality issues that must be considered during processing, interim storage after processing and shipment to a repository?

There is no evidence that DOE has addressed the shipping question. Given that the packages must meet 10 CFR 71 requirements including accident testing, quality assurance plans, transport configurations and criticality evaluations that must be based on different assumptions than those required by 10 CFR 60, it is prudent that DOE consider the many aspects of shipping before canister designs are set.

There is similarly no mention of interim storage after processing; however, if the waste is to be stored in the repository canisters and the canisters are safe in the repository then it is reasonable to assume that the canisters are acceptable for interim storage. DOE needs to articulate this assumption if it is the basis for interim storage.

The Savannah River status report has a one-paragraph criticality statement (5.2.3.3), which acknowledges that criticality stability still must be explored for the melt-dilute option; this is a start but there is insufficient information to judge the adequacy of criticality considerations during processing.

It is my judgment that the transportation issues and justification that the canisters with their contents meet shipping requirements are the most important issue not addressed by DOE at this point. Road accidents might be more limiting than long term survival in a repository. A criticality accident on the road would be much more visible and could potentially have a greater health/environmental impact than a criticality event in the repository. To these ends, DOE must assure that canisters are acceptable not for only the repository, but also for interim storage as well as for transportation.

The Savannah River status report (2.b above) makes a somewhat disturbing statement on page 4.52 in section 4.5.1. That statement is: "The computer codes that we use for criticality calculations for disposal will require benchmarking and/or validating the code." Without validated computer codes and the cross section sets used with those codes, there is no basis on which to proceed with defensible safety evaluations. Also it must be noted that the status report does not identify cross sections being used with the two identified codes. The validity of the cross section sets used may become an issue in dealing with some of the exotic materials being considered (especially the wide range of neutron poisons under study).

In response to the second specific question: Do any of the waste forms produced by the alternatives pose significant internal or external criticality hazards in a repository?

Information presented by Peter Gottlieb as well as information in the Savannah River status report indicates that the current envisioned canister designs will require neutron poisons for direct disposal as well as for codisposal with both High Enriched U-Al fuel (e.g., Massachusetts Institute of Technology (MIT)) and for Low Enriched U-Si-Al fuel (e.g., ORR). The actual neutron poisons to be used are still to be determined; however, several candidate materials have been identified. It appears that a stronger neutron poison will be required (e.g., Gd for high enriched fuel) than for low enriched fuel (e.g., borated stainless steel). The relative dilution rates of fuel versus neutron poison over a long period pose a vexing problem, given that the actual neutron material is still to be selected. Adding to this, the potential cross section validation problems with certain neutron poison materials as well as the quality assurance problem of misloading a canister with the wrong poison or the wrong type of fuel in a given canister, raise the question of the desirability of this alternative over the melt-dilute alternative. The melt-dilute alternative can be designed to remove any requirement for neutron poison material and likewise render relative dilution rate problems easier to define and defend. In my opinion, required use of a neutron poison material with both the direct disposal and the codisposal alternatives represents a significant criticality hazard for the repository—a hazard that can be eliminated by use of the melt-dilute alternative. The dilute-melt will pose additional consideration for criticality during processing; however, the processing system can be designed with positive and monitored control. The internal canister basket would lend itself to a design with easier quality assurance requirements.

It is my opinion, based upon the studies completed to date, that the melt-dilute alternative poses less of a criticality hazard in both shipping and the repository than does either direct disposal or codisposal. Press and dilute/poison was mentioned as another highly attractive alternative; however, almost no information was presented and intuitively this alternative appears to be less attractive from a criticality perspective than melt and dilute. Press and dilute would probably be preferred to either direct disposal or codisposal because it could be carried forward without need of a neutron poison. None of the other nine alternatives were considered as part of this review.

Topic: Cost and Schedule
Consultant: M.W. Angvall

The Principal Investigator posed six questions concerning cost and schedule. These questions and the corresponding responses follow. Information was gathered from various references as well as briefings and meetings held in Augusta on December 2 and 3, 1997, and from revised cost data provided after the December 2 and 3, 1997, meetings.

Question 1: Are the cost data provided by DOE reasonably complete and transparent?

Response: The revised costs provided by DOE appear to be complete. The cost estimates were constructed using the major cost drivers, which together make up the full costs of SNF handling, conditioning, packaging, storage, and disposal for each treatment technology. The estimates were logically constructed using scaling factors, inflation, and project contingencies in a reasonably judicious manner. Financing costs were added as were IAEA implementation costs and NRC licensing costs. Upon reviewing the backup provided in the revised cost study,[1] we can say the costs are reasonably transparent.

Question 2: Are the cost and schedule estimates developed by DOE for the alternative processing options suitable as a basis for comparison and selection of one or more preferred alternatives?

Response: The cost estimates are suitable as a basis for comparison and selection of one or more alternatives. The schedules have been upgraded to reflect reasonably realistic dates and would appear to support the detail work required to further refine the selected technologies.

---

[1] Krupa, J. F. and Carter, J. M., *Savannah River Site Aluminum-Clad Spent Nuclear Fuel Alternative Cost Study, Rev. 1(U)*.

Question 3: Are the cost and schedule estimates developed by DOE for the alternative processing options suitable for budget planning purposes?

Response: The cost and schedule estimates developed for the alternative processing options are not suitable for budget planning purposes. The transfer, storage, and treatment facility cost estimates for the direct co-disposal facility and for the melt and dilute facility were prepared based upon a preconceptual design estimate. However, the estimates for the other technologies as well as the remaining cost factors are of parametric or rough order-of-magnitude quality and cannot be considered accurate forecasts of actual financial requirements.

The schedule estimates are based upon assumptions as to delivery of aluminum-based SNF shipping casks and aluminum-based SNF assemblies to Savannah River; upon projected dates at which the various technologies could be available using a privatization approach (which to date has not been successful) for the transfer, storage, and treatment facility costs; and upon the date on which the repository will be ready to accept shipments. Any significant slippage in any one of the assumed dates could have major cost ramifications for all of the technologies and could affect some technologies more adversely than others.

However, the revised schedules are probably sufficiently realistic to be used to develop estimates for budget planning purposes.

Question 4: Has DOE considered the costs of program delays in its budget development or budget planning for this program?

Response: It would appear that no costs for program delays have been included in the cost estimates. All of the estimates have been based upon the revised dates for shipment to Savannah River, transfer within Savannah River, funding, start-up of new treatment facility operations, and shipment to the repository. Delays in any of the assumed dates will have a negative cost effect on the estimates. Project contingencies were assigned to quantify the uncertainty associated with the implementation of each SNF Technology. This contingency addresses such things as equipment unknowns, complexity of process, and process integration.[2] No program delay costs were included.

Question 5: Are the costs and schedule estimates for implementing the alternative processing options consistent with DOE procedures and systems? If not, has DOE identified what changes must be made to achieve its cost and schedule targets?

Response: The original schedule used in the Task Force Report was directed by DOE in the task force ground rules. This schedule did not permit the development of budget quality estimates. The revised schedules discussed at the Augusta meeting and those listed in the revised cost study would appear to allow development of estimates meeting DOE procedures. DOE appears to have made a policy decision concerning new Savannah River spent fuel management projects that has allowed development of schedules consistent with DOE procedures and systems.

Question 6: Are the cost and schedule milestones that are laid out in the Research Reactor Task Force Report for selecting and implementing an alternative processing option being met?

Response: The schedule milestones presented in the Research Reactor Task Force Report are not being met. However, the schedules have been revised to more realistic dates, which have the possibility of being attained.[3] The question remains as to the dates selected being sufficiently realistic considering all the basic assumptions made, and the number of policy-making parties involved.

---

[2] Krupa, J. F. and Carter, J. M., *Savannah River Site Aluminum-Clad Spent Nuclear Fuel Alternative Cost Study, Rev. 1(U)*, p. 21.

[3] Krupa, J. F. and Carter, J. M., *Savannah River Site Aluminum-Clad Spent Nuclear Fuel Alternative Cost Study, Rev.1 (U)*, p. 3.

Topic: Regulatory and Waste Acceptance
Consultant: Robert M. Bernero

Introduction

The Department of Energy (DOE) is considering alternatives for the final disposition of aluminum-based spent nuclear fuel recovered from reactors here and abroad. The fuel, which contains high-enriched uranium (i.e., approximately 20-93% ^{235}U), is of particular interest because of the potential for diversion and the higher expected rate of corrosion for aluminum.

The National Research Council (NRC) of the National Academy of Sciences (NAS) was asked by DOE to perform a review of DOE's aluminum-based spent nuclear fuel disposition technologies. The evaluation would include the following: (1) examination of the set of technologies chosen by DOE and identification of other alternatives that DOE might consider; (2) an examination of the waste-package performance criteria developed by DOE to meet anticipated waste acceptance criteria for disposal of aluminum-based spent fuel and identification of other factors that DOE might consider; and (3) to the extent possible given the schedule for this project, an assessment of the cost and timing aspects associated with implementation of each spent nuclear fuel disposition technology.

This study was undertaken using the Principal Investigator project model. The analysis here is by one supporting contributor evaluating the regulatory and waste acceptance approaches for final disposition of the aluminum-based spent nuclear fuel.

Background

Until the late 1970s the prevailing concept for final disposal of all high-level nuclear waste was that all spent fuel would be reprocessed and the resulting high-level waste stream would be somehow solidified to prepare the waste for long term disposal. The solidified waste, after reprocessing, would contain only small amounts of fissile isotopes, leaving the way open for specification of waste form performance criteria as the first line of protection against release and transport of the waste isotopes. In 1970 the Code of Federal Regulations, in 10 CFR Part 50,

Appendix F, defined high-level liquid radioactive wastes as "those aqueous wastes resulting from the operation of the first cycle solvent extraction system, or equivalent, and the concentrated wastes from subsequent extraction cycles, or equivalent, in a facility for reprocessing irradiated reactor fuels." That regulation also limited a reprocessing plant's inventory of high-level liquid radioactive waste to that produced in the prior 5 years. That regulation also required shipment of the high-level radioactive waste to a Federal repository no more than 10 years after the waste is separated from the fuel.

In the winter of 1976/1977, Presidents Ford and Carter decided that commercial reactor spent fuel would not be reprocessed, but would be disposed of directly as high-level waste. That decision for the once-through fuel cycle immediately changed the standard form of high-level waste, from waste solidified to meet a performance specification to the as-removed form of commercial spent fuel, covering a wide range of sizes, burnup levels, etc.

Later, the U.S. Nuclear Regulatory Commission (USNRC) promulgated regulations, 10 CFR Part 60, to require that these high-level wastes be emplaced in waste packages providing "substantially complete containment" for 300-1,000 years, in an engineered barrier system that releases less than 1 part in 100,000 after 1,000 years, in a geologic medium with a groundwater travel time of at least 1,000 years (10 CFR Part 60.113). In addition, in Part 60.112, the USNRC regulation requires the repository to meet the performance standards of the generally applicable environmental standard set by the Environmental Protection Agency (EPA).

The original final disposal standard from the EPA, 40 CFR Part 191, set quantities of permitted release, by isotope, assessed at 10,000 years. These permitted release quantities are associated with health effects imputed to collective population doses. By the Energy Policy Act of 1992 (P.L. 102-486) the Congress set aside that EPA standard for Yucca Mountain, the first candidate high-level waste repository under investigation. Instead of that standard, Section 801 of the Act directed EPA to promulgate standards to ensure protection of public health from high-level radioactive wastes in a deep geologic repository that might be built under Yucca Mountain in Nevada. By this provision, EPA must set the standards to ensure protection of the health of individual members of the public. The standards will apply only to the Yucca Mountain site. To assist EPA in this endeavor, Congress asked the National Academy of Sciences to advise EPA on the technical bases for such standards.

The National Research Council Board on Radioactive Waste Management formed a Committee on Technical Bases for Yucca Mountain Standards. That Committee has issued its detailed report, *Technical Bases for Yucca Mountain Standards* (National Research Council, National Academy Press, 1995). The EPA standard 40 CFR Part 191 has since been revised and reissued applicable to the Waste Isolation Pilot Plant (WIPP) and high-level waste repositories other than Yucca Mountain, but EPA has not yet issued a revised Yucca Mountain standard.

Regulatory Approach for Yucca Mountain

Yucca Mountain is being investigated as a possible repository for high-level radioactive waste under the terms of the Nuclear Waste Policy Act of 1982 and the Nuclear Waste Policy Amendments Act of 1987. The repository, if approved, would be used principally for the disposal of commercial reactor spent fuel, with a statutory limit of 70,000 tons (from the 1987 Act). A decision by President Reagan permitted some of this capacity to be used for defense high-level wastes. It is assumed here that disposal of the aluminum-based spent fuel would be in Yucca Mountain, co-disposed with glass logs from defense high-level wastes. Therefore, the regulatory approach and waste acceptance basis for aluminum-based spent nuclear fuel are the same as those for Yucca Mountain.

The USNRC regulation, 10 CFR Part 60, promulgates the phased regulatory approach required by the Nuclear Waste Policy Act of 1982. That approach consists of several phases: site characterization, license application (after Presidential notice to and approval by the Congress), construction authorization, license to receive waste, and finally license amendment for closure. At the present time, Yucca Mountain is still in the site characterization phase, with DOE acting on a Site Characterization Plan first submitted to USNRC in late 1988. That Site Characterization Plan was reviewed by USNRC following the procedures specified in 10 CFR Part 60.18, *Review of Site Characterization Activities*. By those procedures the USNRC does not approve the activities or thereby commit to issuance of a license. Rather, the USNRC reviews the activities, raising questions, providing comments, or even raising objections in matters that appear to be so seriously deficient as to raise doubt about the possibility of future licensing. Thus, the first Yucca Mountain Site Characterization Plan was reviewed and received many comments with two objections, that the quality assurance program for data gathering was inadequate and that the repository design control process was not coherent (letter R. M. Bernero, USNRC, to S. M. Rousso, DOE-RW, dated July 31, 1989). Those objections and major comments have long since been resolved and the site characterization phase has continued with a clear public record of the characterization, continuing analysis, and review. Following the procedures of characterization phase review the USNRC does not specifically approve any data or analysis but does document its review thereof to reveal any objections, comments, or questions. The DOE is then able to address and resolve them with USNRC review and acknowledgment, but without explicit USNRC approval.

Until the EPA has acted to promulgate a new performance standard for Yucca Mountain disposal, the DOE approach for performance assessment of Yucca Mountain is to conduct total system performance assessment, using state of the art methods, assessing performance against the standards likely to result, considering the advice of the Committee on Technical Bases for Yucca Mountain Standards. For example, performance assessments are not stopped at 10,000 years but are carried out over the Committee-recommended 1 million years (which indicate that peak projections appear beyond 10,000 years, at several tens of thousands of years).

In another area of significance here, DOE has established a process for determining the acceptability of candidate forms of high-level radioactive waste. That process is laid out in *DOE/RW-0351P Waste Acceptance System Requirements Document,* December 1996. An important element of determining waste acceptability is the conduct of a performance assessment to determine that disposal of this particular candidate waste will not significantly affect the overall performance of the repository. In addition, if the candidate waste, like spent nuclear fuel, contains fissile material, the performance assessment must include a criticality safety analysis. The December workshop on this project was told how DOE is now systematically analyzing the various forms of spent fuel in DOE possession, including the aluminum-based spent fuel, in this way. Consistent with the site characterization phase of Yucca Mountain activities, DOE is conducting this work with review and comment by USNRC, not seeking approval.

Evaluation of DOE Aluminum-Based Spent Fuel Alternatives

The evaluation of the alternatives being considered by DOE for management of aluminum-based spent nuclear fuel began with four questions:

• Has DOE-Savannah River identified the appropriate criteria for aluminum-based spent fuel from the draft waste acceptance criteria document that has been prepared by the Yucca Mountain program?
• Which of the waste forms is likely to be the most acceptable for disposal in a repository relative to commercial spent fuel and vitrified logs? For those waste forms that are unlikely to be acceptable, has DOE considered alternate processing options?
• Are the waste acceptance criteria that have been identified by DOE suitable for selecting among the alternative processing options?
• Is DOE-Savannah River making an adequate effort to stay current with changes in waste acceptance criteria?

The evaluation here addresses each of these questions in order, beginning with the first. DOE-Savannah River has not attempted to identify the appropriate criteria for aluminum-based spent fuel acceptance unilaterally. Rather, it has agreed with and supported the expert team in the service of DOE's Office of Civilian Radioactive Waste Management (DOE-RW) to conduct the performance assessments needed for the waste acceptance process described in DOE/RW-0351P. That analysis is also available for review by the USNRC. Thus, DOE-Savannah River has identified the appropriate process for determining waste form acceptance and joined in that process.

In considering the second question, it should be noted that there is some ambiguity in the DOE approach to simply reprocessing the aluminum-based spent fuel using the existing facilities at the Savannah River Site. The recovered fissile material could join other high-enriched uranium streams in the DOE system, and the waste liquid could pass to the vitrification process to become ordinary glass logs, already established as a waste form. Instead of following this option, the DOE-RW and DOE-Savannah River waste acceptance process team has conducted detailed analysis of only one alternative, direct co-disposal. Direct co-disposal is the option where the intact spent fuel is loaded in a suitable canister, small enough in size to fit in the center of the proposed disposal array of five glass logs in a ring within a large canister. The direct co-disposal is evidently more attractive than the base case of direct disposal, because it uses the otherwise unused space in the center of the glass log canister and thereby obtains a substantial degree of self-protection from the surrounding waste logs. Direct co-disposal is also the evident limiting case for waste acceptance since it brings the unmodified corrosion characteristics and the higher nuclear reactivity of this spent fuel directly into the repository. The approach seems to be that if the direct co-disposal option yields an acceptable waste form, there is no need for further consideration of waste acceptance in weighing alternatives. That approach is reasonable as revealed by the consideration of the next question.

The third question now simplifies to whether direct co-disposal is an acceptable waste form. The results of analysis so far obtained by the DOE-RW/DOE-SR team were presented at the December workshop. The team appears to be using appropriately conservative models to explore the alternative scenarios of corrosion and material slumping in order to discern the possibility of nuclear criticality. It should be noted that the team analyzed two types of this spent fuel as bounding, the fuel from the MIT Reactor and that from the Oak Ridge Reactor, each with its own tailored canister. Careful choice of and arrangement of poison in the canister for each specific fuel is important to obtain satisfactory results. The analytical results indicate that in situ criticality is highly unlikely, and we were told that if one did occur it would be a low yield, Oklo-type event, not a prompt criticality as suggested by some recent authors. The acceptability of these nuclear reactivity conclusions was linked to the expectation that the relevant USNRC regulation, 10 CFR Part 60.131(h), would be changed by USNRC rulemaking before repository licensing. This is a reasonable assumption since that regulation is a standard deterministic, double contingency requirement typically applied to fissile material handling. It is my understanding that USNRC recognizes and agrees with the need to change this regulation to a probabilistic, performance assessment model for the repository.

With the apparently acceptable nuclear reactivity conclusions, the team presented its current analysis results for all waste releases from the repository over a million year time period and the aluminum-based spent fuel contribution to those releases. The releases from the aluminum-based spent fuel are not significant; they are several orders of magnitude lower than the repository overall. Therefore, the results so far indicate that direct co-disposal is an acceptable waste form and any other option, if chosen, could yield an acceptable waste form.

The fourth question really addresses where DOE is going from here. It is my impression, based on the presentations and discussions at the workshop, that reprocessing, direct co-disposal, or a combination of the two will be the options of choice. If direct co-disposal is pursued, the analyses should be completed and documented, especially the parts related to specific canister design and nuclear criticality analysis. The USNRC should be asked for detailed review of the analysis and for a commitment to the rulemaking to change 10 CFR Part 60.131(h).

Summary

DOE-Savannah River is following a reasonable regulatory approach for establishing the acceptability of waste disposal forms for aluminum-based spent nuclear fuel. It has nearly completed substantial analysis to demonstrate the waste acceptability of the bounding and most promising option, direct co-disposal. The work is not yet complete but the path to completion is clear.

Topic: Processing and Remote Handling Considerations
Consultant: Joseph Byrd

Introduction and Background

Background and details of the aluminum-Based Spent Fuel Program were presented in a plenary session on 2 December. Materials that were presented are summarized below.

In consideration of a 1992 DOE decision to phaseout reprocessing activities throughout the DOE complex, a program was initiated to develop non-reprocessing technologies to be ready for implementation by 2000. DOE assembled a Task Team in November 1995 to develop a technical strategy for safe and economical handling, treatment, and disposal of aluminum-based SNF in hand and to be received. Eight non-reprocessing options along with the reference processing case were selected and studied. The options included direct disposal, co-disposal, dilution, and advanced treatments. The findings and conclusions of this Task Team were consensus-based using four weighted evaluation criteria: confidence in success, cost, technical suitability, and timeliness. Direct/co-disposal and melt/dilute options were recommended as alternatives to reprocessing.

An Alternate Technology Program was initiated to move forward with development of the options recommended by the Task Team. Objectives and deliverables have been set through FY-98. All eight of the non-reprocessing options were discussed including reviews of waste forms, waste containers, and handling/processing issues. Results of the studies for metallurgy and corrosion issues and disposal criticality analysis were presented and discussed.

Processing/Remote Handling Issues

On December 3, small group breakout discussions were held to discuss details and remaining issues on specific topics. The groups concentrated on the topics specified in the technical expert assignments for the workshop.

The NRC project's principal investigator provided the following questions to be addressed for the processing/remote handling considerations. These specific questions and other issues were discussed in the breakout session with knowledgeable meeting participants including representatives from Savannah River Site, Argonne National Laboratory, and Idaho National Laboratory. Each of the eight technology options was discussed relative to the following questions. The participants were able to provide in-depth discussions relative to the following questions. The following answers to the questions are based on those discussions.

1. For each of the processing options evaluated by DOE, are the processing steps used as the basis for assessment and comparison (other than direct co-disposal) technically credible? That is, are they likely to work as described and produce the products and results assumed?

Yes. The Task Team did a good job in its evaluation of the options. It considered a reasonable set of criteria on which to base its decisions and recommendations. Some options were eliminated and some added during the evaluation process. The Task Team was diverse.

2. Are there other processing options that should be considered by DOE for disposition of aluminum-based spent fuel?

No other options that should have been considered are apparent to this reviewer.

3. Do the inner container designs appear adequate to contain the waste forms resulting from the various processing options? Are they over-designed for the intended application?

The inner containers appear to be reasonable approaches for containment of the waste from various processing options.

4. Are DOE's basic material handling plans, pool use, and other facility needs reasonable, and are remote handling technologies available to meet these needs?

Material handling for pools and repositories is not an issue. There are some remote handling issues related to some of the process options.

Remote handling systems for all options would require some systems development (i.e. systems are made up of standard components but are not "off the shelf" items). Remote handling equipment would require more development for the handling laminates for the Press/Dilute option, but proven technologies are available to meet this need. Significant remote handling problems are anticipated for the Plasma Arc Treatment option. More design information would need to be available before remote handling assessments could be made. No major problems are anticipated for the remaining options. Technologies are available to meet the remote handling needs. However, more design information would have to be available for all options before an assessment could be made on exactly how much remote handling system development would be required.

A preliminary design has been done for a Transportation and Storage Facility that would accommodate any of the process options. The material handling and remote handling are assumed to be the same for getting materials into and out of whatever process is selected. This is a reasonable assumption when considering those processes most likely to be selected. Some chopping and material conditioning will be unique to the individual process and must be addressed in the development of those options (Plasma Arc Treatment, Chop/Dilute, and GMODS). The preliminary facility design uses proven methods for manipulators and overhead cranes. An electric cart system is used for transportation within the Facility. For the final design more attention should be given to remote handling. State-of-the-art Automated Guided Vehicles (AGVs) should be evaluated. Also, enhancements to remote handling are available to provide more effective and efficient operations: for example, graphical operator interfaces, simulator operation modes, telerobotic operations (combined teleoperator and semi-autonomous modes), and swing-free cranes. These technology enhancements should be considered while keeping in mind the overall objectives of efficiency and safety. Databases and database management should be incorporated to inventory and track materials through the entire process.

5. Are the technical requirements of the various alternative processing options sufficiently well defined so that reasonable judgments can be made about the likelihood of success of implementing them?

Yes. The Task Team in its analyses adequately considered technical requirements.

6. Are there large differences in likelihood of success of implementing the various processing options?

In general, some uncertainties such as the percentage of uranium that will be allowed will have an impact on most options. Lower percentage requirements will result in higher temperatures, off-gas problems, etc.

The GMODS (Glass Material Oxidation and Dissolution System) and Plasma Arc Treatment options have more unknowns and would require significantly more development in order to confirm their likelihood of success. The Plasma Arc Treatment has anticipated high temperature problems and complex issues related to the shredder mechanism and process. The GMODS has been proved on bench scale mockup. However, lead in the off-gas and the shear mechanism/process into the melter will present problems.

Handling and characterization will be problems with the Press/Dilute Option. More design work would be required.

General Issues

The following are general concerns about issues related to the project.

• Political decisions appear to be the project drivers that override any technical considerations beginning with the 1992 decision to phase out all reprocessing operations throughout the weapons complex. This may make sense for consideration of new major facilities. However, this decision has even affected the consideration of using an already existing facility, H-canyon at SRS, an option that would be most cost-effective.

• Nonproliferation issues appear to be overemphasized for this problem. Unknown international "perceptions" appear to be serious concerns, although the research reactors' SNF handling, processing, and disposal are only the "tip of the iceberg" of the national waste problems. This appears to be a very small nonprofliferation issue compared with the overall nonproliferation issues in total waste management, waste disposal, and waste storage. Emphases on past examples do not seem appropriate or necessary for this problem.

• Remote handling systems development costs have not been factored into the costs associated with the technology options. This probably does not substantially alter the total costs for those options under consideration but may change the ranking since many of the cost estimates were very close together.

Conclusions and Recommendations

1. Advanced technologies in remote handling. No major problems associated with remote handling for the options under consideration are anticipated. However, there were no indications that the latest technologies in remote handling were being considered in the facilities and processes designs. Basic manipulators, cranes, and electric carts can be greatly enhanced using proven advanced technologies such as swing-free cranes, telerobotic manipulators, graphical operator interfaces, system simulators, and advanced AGVs (automated guided vehicles). These technologies should be considered and evaluated for all appropriate phases of the process in order to assure the most effective and efficient operations.

2. Computer databases. There were no indications that advanced computerized databases were being considered for inventory and tracking of materials through the processes. It is recognized that the workshop did not specifically address this issue. However, the use of these systems is recommended to enhance security, safety, record keeping, reporting, and efficiency of the entire operation.

3. Uranium content in final product. There is still an undecided issue concerning allowable percentage of uranium in the final products. This issue should be settled as soon as possible so that technical uncertainties associated with this factor can be resolved.

4. Nonproliferation and criticality issues. Many nonproliferation and criticality issues are still unresolved. Since these major issues are option dependent, they should be resolved as soon as possible so that the process can be selected and designs can proceed in order to meet the desired schedule.

5. H-canyon option. Due to earlier political decisions, the use of H-canyon at SRS was not considered as a viable solution to the research reactor spent nuclear fuel problem. The use of H-canyon at SRS appears to be the most cost-effective and practical solution. Since budget issues are major concerns in the overall national environmental waste problem, the decision not to use the existing H-canyon should be reevaluated and reconsidered.

References

1. *Technical Strategy for the Treatment, Packaging, and Disposal of Aluminum-Based Spent Nuclear Fuel: A Report of the Research Reactor Spent Nuclear Fuel Task Team.* Volumes I and II, U. S. Department of Energy, Office of Spent Fuel Management. June 1996.
2. *Alternative Aluminum Spent Nuclear Fuel Treatment Technology Development Status Report.* Savannah River Technology Center, Strategic Materials Technology Department, Materials Technology Section. WSRC-TR-97-00345. October 1997.
3. *Savannah River Site: FY97 Spent Nuclear Fuel Interim Management Plan.* Westinghouse Savannah River Company. October 1996.
4. *Technical Strategy for the Management of INEEL Spent Nuclear Fuel: A Report of the INEEL Spent Nuclear Fuel Task Team. Idaho National Engineering and Environmental Laboratory. March 1997.*

Topic: Metallurgy and Corrosion
Consultant: R.L. Dillon

Introduction

The emphasis in this report is on questions concerning options for the disposition of aluminum spent nuclear fuel (Al-SNF) raised by Milt Levenson, Principal Investigator, and Kevin Crowley, Study Director, in their "Instructions to the Technical Experts" (1). The instructions indicate that focus of the study shall be (a) dry storage of fuel and achievement of a road-ready status for the fuel package, (b) processing and preparation of optional waste forms and corresponding waste form properties, (c) interactions of waste form with the canister, and (d) review of the status of Co-Disposal and Melt-Dilute/Press-Dilute disposal options.

The reference material made available to the review panel was researched and written up by a small team of scientists and engineers who appear as authors in various combinations. A given piece of work may be covered several times in various status and topical reports. In view of the interlocking authorship and repetitive reporting I have not felt it necessary to cite all the documents in which a given piece of information appears.

Question 1a: Are the DOE plans for fuel handling, drying, and interim storage technically credible? Are these steps adequate to prevent significant fuel corrosion?

Response: (1) Dry storage criteria for Al-SNF require the fuel remain in a condition mechanically suitable for "full safe retrievability" (2) for manipulations such as fuel rearrangement within a container, movement of fuel between canisters, adjustment of the gas environment, or withdrawal of fuel assemblies for examination or testing. Corrosion limits must assure that residual cladding has the requisite strength. In addition, processes that distort the clad-fuel material interface by blistering must also be avoided.

These requirements have led to procedures that limit the reactant inventory within the container, after drying, to amounts that do not significantly reduce effective cladding wall thickness. Limiting maximum cladding-fuel temperature to less than 200 °C in dry storage controls fuel blistering and creep (3, Sec 3.2).

The effectiveness of the outgassing during the drying process has been shown, but there may still something to learn from tests on irradiated MTR fuel (3). Such tests are dependent on the availability of the Instrumented, Shielded Test Canister System (3). The response to the first part of questions 1 is that observance of corrosion limits during storage and drying and access to the Shielded Test Canister System for validation purposes will permit handling even with pitted fuel (2).

Question 1b: Are these steps adequate to prevent significant fuel corrosion?

Response: Laboratory studies have measured the effects of drying on residual moisture in a test chamber. Scaling to a fuel canister, quantitative consumption of the after-drying water vapor inventory does not approach corrosion or hydrogen limits. These extrapolations need validation that the Shielded Test Canister System will make possible. Studies performed at INEEL and Hanford have shown that mechanical pumps can outgas fuel and large nuclear components to water levels acceptable for dry storage of Al-SNF (4).

Question 2: For each of the processing options evaluated by DOE, are the processing steps used as the basis for the assessment and comparison technically credible? Are they likely to work as described?

Response: It is proposed here to consider only four of the eight options: Direct/Co-Disposal (3, Sec 1.1); Press and Dilute (5, App F), or Melt-Dilute (3, Sec 5.1), and the Electrometallurgical Option (5, App F). I have taken the cue from the Research Reactor Spent Nuclear Fuel Task Team who indicate in reference 3, Sec 6.2, that Direct/Co-Disposal may be regarded as the primary approach with Press and Dilute or Melt and Dilute as backup. It also recommend electrometallurgy as a secondary and diverse option. This report considers only these technologies.

It is not the object in this discussion of processing options to undertake a general review. Comments are confined to four of the eight options and to the corrosion implications of the process itself and of the waste products that emerge from the processing.

• Direct/Co-Disposal Treatment (3, Sec 4.0): This process proposes to take fuel that has met criteria relating to its condition (cladding integrity, pit depth and diameter and, oxide thickness as it exists in a basin) and place it with other waste forms in an efficient array within a container. After outgassing to 4.6 torr to remove free water, the container is sealed for placement with other waste forms (e.g., glass logs in a canister) for eventual shipment to the repository. What is expected of the Direct/Co-Disposal Option is a method of treating Al-SNF using current technology, that will provide a waste form for safe interim storage, and will be road ready when regulatory approvals are obtained for shipment to the repository. The methodology depends on proven techniques or those sufficiently developed to be confidently applied to the waste form package (e.g., fuel drying), all at minimum cost. There are questions regarding the admissibility of the Direct/Co-Disposal product to the repository that require resolution.

Regulatory problems regarding waste form admissibility to the repository are matters outside the scope of our charge. Beyond that, the technology is in place, and I see no corrosion reason not to put the process in operation.

• The Direct/Co-Disposal waste treatment process does not stand alone. The Melt-Dilute (3, Sec 5.0) option offers some superior features, but it involves technologies proven only on a reduced scale. In this option, fuel assemblies are melted with additions of depleted uranium for isotopic dilution and additions of aluminum—and, where needed, depleted uranium—for a desired aluminum-uranium (Al-U) alloy composition. One advantage of the M-D approach is that it permits isotopic dilution if that is required. There are process options that remain to be considered with respect to melt pouring temperatures and methods and melt composition, which can improve the cast product for specific applications. Sampling during processing of the molten, homogenized pool of several fuel assemblies at once may provide superior and less expensive analytical data. The cast product is compact relative to intact fuel assemblies, with the potential to reduce the number of canisters that must be provided for in the repository. In principle, M-D could deal with uranium oxides and silicides as well as Al-U alloys. However, difficulties are encountered in alloying UO_2, which is a more stable than Al_2O_3 at higher melt temperatures.

Three test methods have been selected for assessment of SNF degradation and the release of radionuclides from M-D product (3, Sec 6.1). However, the greatest value for these techniques is likely to come in studying reactions between canister materials and repository waters, or fuel materials and repository waters. This is not an environment we are to consider. For evaluation of these test procedures the test matrix proposed by Savannah River is needed (3, Sec 6.0-6.25).

• Press-Dilute Treatment (5, App F): This is a waste volume compaction process with isotopic dilution. Dilution is accomplished by layering the fuel material with depleted uranium and pressing to form a block. Six pressed blocks are placed in a stainless steel can and welded shut. This is a simple technology that could be put into operation with relative ease. No special corrosion problems are foreseen.

• The Electrometallurgical treatment of Al-SNF (5, App F) is a two stage process plus a head-end facility. The Al-SNF is melted and cast into anodes for the electrorefiner. The first electrorefining step is the transport of aluminum from the Al-SNF anode in a LiF-KF electrolyte to the electrorefiner cathode, leaving uranium and metallic fission products behind in the basket of the anode. The aluminum is disposable as low-level waste. The contents of the anode basket are transferred to a second electrorefiner where high purity uranium is recovered. Fission products are converted to oxides that provide a feed stock for a glass melter. Fission products xenon and krypton are released to the inerted enclosure; other volatile fission products are trapped.

High throughput of electrorefined uranium has been demonstrated at ANL. High throughput electrorefining of aluminum has yet to be shown, along with scrubbing alkaline earth products from the aluminum electrorefiner salt. Design of an engineering-scale electrorefiner is underway. Nothing is said about corrosion during electrorefining. Until more is known about the feasibility of the process it is probably too early to give corrosion much priority.

The benefits of a functional electrometallurical process are the recovery of enriched uranium for potential reuse and the consequent reduction of nuclear waste storage canisters requiring disposal. If reuse of the recovered enriched uranium is not an option, then the EM product has few apparent advantages over the Melt-Dilute Process.

Question 3: Since the amount of water in the repository is likely to be somewhat limited, would filling the canister void space with aluminum or some other sacrificial material make any difference in the long-term corrosion of the waste form?

Response: The drying operation is designed to remove free water in the container prior to its sealing off and isolation. Through the drying operation, residual water is limited to amounts consistent with an aluminum corrosion allowance of 0.3 mil. and a hydrogen gas limit of less than 4 percent by volume, the lower limit of hydrogen gas flammability (3). Void-space pressurization and the limits on corrosion product hydrogen generation are more restrictive than the corrosion limit (2,3). The water that will remain in the void space after water vapor levels are reduced to a few torr in the drying operation will be insufficient to endanger either the corrosion limit or the void space gas pressure or composition limits. In "dry storage," without some inexplicable entry of water, the addition of aluminum to the void space is largely without consequence except to the extent that the reduced free volume would increase the gas space pressure for a given amount of corrosion.

This question is more meaningful in the context of repository behavior, which is outside our charge. It is most plausible to me that water for entry into the waste container fuel material will first have penetrated the massive carbon steel overpack and then the inner canister liner. The possibility of crevice corrosion in the annulus between the inner barrier (alloy 625, Ref. 6) and the outer barrier (A516, Ref. 6) of the canister will require assessment. Access of considerable water to the repository is implied. If this scenario is rational, the addition of sacrificial aluminum to the canister void space would be inconsequential.

Question 4: Will any waste forms resulting from any of the alternative processing options be likely to increase internal corrosion of a standard repository container compared to spent commercial fuel or vitrified glass logs? That is, are there likely to be interactions between the waste form and the inner container in excess of what would be expected for commercial fuel or vitrified glass logs.

Response: *Direct/Codisposal.* Aluminum clad fuel removed from pool storage for direct disposal is expected to bring with it a small amount of water along with any deposits from the reactor coolant or pool storage. Treatment in a tumbling washer will remove the bulk of the crud and loose deposits (4). After vacuum-drying, residual water levels for encapsulated fuel assemblies are too low to challenge limits set for incremental corrosion of Al-clad in the fuel storage container. Gas pressure (water vapor or H_2 or a combination of those) in the vacuum-dried waste container will be well under acceptable pressure limits. There is no reason to suspect nonvolatile corrosive substances to be introduced to the container by way of the spent fuel charge.

  Press and Dilute. Vacuum drying is applicable to press and dilute Al-SNF processing. Similar assurances regarding water vapor and nonvolatile corrosive substances are justified.

  Melt and Dilute. This Al-SNF treatment process generates a melt of uranium, aluminum, transuranics, and fission products cast compactly into a mold for disposal in a waste container. Given the high temperature of processing, the possibility of release of water or corrosive impurities to the container from the waste form is further reduced over the processes for direct/codisposal or press and dilute treatment.

  Electrometallurgical Processing. This process also involves melting of the waste form and negligible likelihood of introducing water or corrosive species to the container environment.

Question 5: What is the status of R&D activities at Savannah River on the Melt-and-Dilute and Co-Disposal options? Are the R&D activities appropriately focused and are they likely to lead to useful outcomes?

Response: An instrumented canister was designed and fabricated to validate the drying and storage criteria for a road-ready container. The instrumented test chamber will accept an Materials and Test Reactor (MTR) fuel assembly in a chamber instrumented to measure and record the temperature of the fuel cladding, the ambient gas temperature, the gas species present, relative humidity, and windows to determine visually the condition of fuel material surfaces. The instrumented canister is suitable for corrosion measurements as determined by water consumption and hydrogen generation. Alternatively, drying of a fuel assembly can be followed as a function of temperature, relative humidity, and time. The instrumented canister is a necessary capability and validates more oblique laboratory studies to the same purpose for the corrosion and drying studies of irradiated fuel (3, Sec 4.2).

In the corrosion area, work done and the proposed future efforts are generally viable and support proposed handling plans. The corrosion models and the range over which they apply are examined later at the conclusion of question responses.

There are several items which have not been treated explicitly in the literature handouts that may deserve some consideration:

• Question 4 above points up the lack of information concerning the choice of material for the waste container and its interactions with the waste form. I refer here to the can into which the waste is placed, whether fuel assemblies, or castings or molten metal from the Melt-Dilute process. Certainly the can could serve better as a barrier to water intrusion if the material and its metallurgical condition were wisely chosen. There is casual reference to alloys XM19 and 316L SS (6) but without comment or any indication of how these materials were selected. Melt-can interactions may be understood, but they are not discussed.

• Radiation reacts with water vapor to generate nitric acid. Both radiation and dilute nitric acid have been shown to accelerate aluminum corrosion rates in vapor phase reactions. Possibly the radiation effect on aluminum corrosion can be accounted for by nitric acid generation. Such is the implication of the parallel corrosion studies: on one hand, using a solution of one part concentrated nitric acid and 6 parts of water and, on the other, water and a gamma source of about 1.8×10^6 rads/hr. It is not clear otherwise why that particular acid solution was chosen for comparison with the gamma irradiated solution. The equilibrium of nitric acid, air and water vapor, and gamma radiation is not discussed. A better understanding of radiation effects requires such a parametric analysis.

• Here are some comments about the corrosion model that come out of my involvement with aluminum corrosion some years ago. Corrosion is strongly influenced by the aluminum surface temperature.

The storage temperature, like all storage parameters, is based on the requirement that Al-SNF must be safely retrievable during treatment and subsequent storage. The interim storage temperature limit has been identified as 200 °C. To keep the concentration of corrosion product hydrogen below the combustion limit it must constitute less than 4 percent of the gas volume. The maximum acceptable gas pressure within a waste container is 60 psi. With these limits set, the tolerable amount of reaction between stored fuel and the gas environment can be determined.

The rates and temperature dependence of the interactions between the container environment and the clad and fuel surfaces not only are important in themselves but can lead to a mathematical model helpful in understanding the corrosion mechanism. The principal virtue of a model is the ability to extrapolate to temperatures where data are not at hand. It is very much to the point to know when accelerated corrosion is likely to occur and its approximate rate.

In liquid water environments, cladding alloys also corrode by a parabolic rate process. There are differences among alloys, but for relatively brief exposures at 200 °C or less, corrosion is proportional to the square root of time. For longer periods of time or higher water temperatures the corrosion process "breaks away." At breakaway the rate becomes linear with time and is much accelerated. Hanford work related time to breakaway and breakaway corrosion rates to temperature, both by Arrhenius type expressions. Savannah River has not looked for this particular type of relationship. However I found it very interesting that specimens in high temperature vapor, corroding in the rapid linear breakaway corrosion mode, continued to corrode by the same breakaway mechanism when transferred to a lower temperature vapor environment. In this present work, recognizing that accelerating rates are probable at long exposure or high temperature is probably sufficient information to avoid underestimating corrosion effects whether rates are predicated by the Savannah River or Hanford approach.

• Inclusion of other corrosion variables in the mathematical model may be possible with information already at hand. Corrosion rates have been determined in isothermal vapor environments at several initial relative humidities, which depending on the experiment may or may not decrease with time. Such an inquiry should start at the beginning and establish the order of the water vapor-aluminum reaction. There could be some use for a corrosion model in which water vapor pressure or relative humidity could be directly cranked into the rate expression.

Conclusions

The body of corrosion work presented to the expert committee is quite impressive. The storage basin-related corrosion studies not only are complete but have resulted in pool cleanup procedures that have eliminated fuel pitting in the basin. It is difficult to find much of a corrosion threat during dry storage, but the work that should be done on container drying requirements and the corrosion consequences of the small amount of residual water has been done. Of particular interest to this observer was the characterization of Al-U alloys and the rationalization of the corrosion behavior of these materials. While similar work has been done on the solubility of radionuclides from glass, the work on radioisotope release from Al-SNF material is new information on an entirely different class of materials and of general technical interest. This is in part because of the projected use of the three test protocols newly applied to Al-SNF characterization. Along with these activities, which are presently laboratory tests, are engineering-scale validation studies in which irradiated MTR fuel assemblies can be monitored under dry storage environmental conditions for fuel and gas temperature, gas species present, pressure, relative humidity, and the visual condition of the fuel material.

I found no significant deficiencies in the Savannah River corrosion program. The water basin corrosion effects have been thoroughly studied. There is little reason to expect new phenomena to show up. Corrosion in dry storage environments will be limited by the small and controlled availability of water.

References

1. Levenson, M., and K. Crowley, Study Director, "Instructions to the Technical Experts," November 14, 1997.

—

2. Sindelar, R.L., H.B. Peacock Jr, P.S. Lam, N.C. Iyer, and M.R. Louthan, "Acceptance Criteria for Interim Dry Storage of Aluminum-Alloy Clad Spent Nuclear Fuels," WSRC-TR-95-0347 (U), March 1966.
3. "Alternative Aluminum Spent Nuclear Fuel Treatment Technology Development Report (U)," WSRC-TR-97-00345, October 1997.
4. Large, W.S., R.L., Sindelar, "Review of Drying Methods for Spent Fuel," WSCR-TR-97-0075, April 1997.
5. "Technical Strategy for the Treatment, Packaging and Disposal of Aluminum-Based Spent Nuclear Fuel," Vol. ll.
6. Benton, H.A., "Waste Form and Co-Disposal Waste Package for Aluminum-Based Research Reactor Fuel," NAS Review of Al-Based SNF Alternative Technology Selection, December 2, 1997.
7. Abashian, M.S., "Mined Geological Disposal System Waste Acceptance Criteria," B00000000-01717040600-00095 REV00, September 1997.
8. Dillon, R.L. and V.H. Troutner, "Observations on the Mechanisms and Kinetics of Aqueous Aluminum Corrosion," HW-51849, September 30, 1957.
9. Dillon, R.L., "Observations on the Mechanisms and Kinetics of Aqueous Aluminum Corrosion II," HW-71756, November 1971.

Topic: Cost and Schedule
Consultant: G. Brian Estes

A list of questions relating to cost and schedule issues follows together with my responses. Information was gathered from a review of references and, a series of briefings and interactive meetings held in Augusta, Georgia on December 2 and 3, 1997, and documents that were provided subsequent to the meetings.

Question 1. Are the cost data provided by DOE reasonably complete and transparent?

Response. The cost data contained in the report of the Research Reactor Spent Nuclear Fuel Task Team, *Technical Strategy for the Treatment, Packaging, and Disposal of Aluminum-Based Spent Nuclear Fuel, Vols. I and II*, appear to have been reasonably complete and transparent as of the time they were prepared and appropriate for the level of development of the alternatives considered.

Cost estimates were built in a logical manner, and data from approved studies were used with a reasonable application of scaling factors, uncertainty factors, and adjustments for inflation. In addition, experience from similar or related projects underway has been factored into the estimating factors.[1]

While some elements of cost appear to be excessive (for example, the combination of construction inspection, project support, project management, and construction management totals 23 percent),[2] they are consistently applied and do not affect the cost comparison.

The question of whether the use of existing facilities rather than constructing new was considered was not addressed in Vol. I of the Task Team report. However, Vol. II[3] and briefings in Augusta, Georgia[4] disclosed that preliminary cost estimates showed higher costs for modifying existing facilities. The primary drivers of the higher costs in this case are operation and maintenance costs of the wet basins, and the fact that other existing facilities would need significant upgrades to meet current Nuclear Regulatory Commission standards, which may be required for facilities completed after 2002.[5]

Question 2. Are the cost and schedule estimates developed by DOE for the alternative processing options suitable as a basis for comparison and selection of one or more preferred alternatives?

Response. The cost estimates appear to be adequate to limit the field of candidate technologies to two or three for further refinement prior to final decision. The schedules contained in the report are unrealistic, but this does not appear to affect the final choice of technologies. Accordingly, there appears to be a reasonable basis to proceed with the refinement of three technologies currently being pursued in detail.

The Transfer Facility Project was not included in the FY98 program as listed in the Savannah River Site FY97 Spent Nuclear Fuel Interim Management Plan[6] on which the Task Force Report schedule was based. The original schedule was probably unachievable even with perfect DOE focus and universal support (personal opinion) but was directed by DOE in the Task Force ground rules.[7] A schedule summary presented at the Augusta meeting indicates a three to five year slip in initial operating capability and substantial increases in costs.[8] This affects all technologies, and a review of the time-sensitive cost elements, primarily wet basin operating costs, indicates there would be no change in relative standings among the alternatives. A detailed cost report issued on December 12, 1997, to support an ongoing independent report on non-proliferation issues shows reasonable agreement in proportional costs. For example, life cycle costs for electrometallurgy were 33 percent higher than direct co-disposal and are now 40 percent higher. Likewise the same costs for dissolve and vitrify were 67 percent higher than direct co-disposal and are now 66 percent higher.[9] This latest report further confirms the original Task Force report conclusions on proceeding with a limited number of options.

Credit for sale of commercial grade uranium resulting from processing/co-disposal and electrometallurgy options is listed in the comparisons.[10] During the Augusta meeting it was determined that while a sale has not yet taken place, negotiations are underway with the Tennessee Valley Authority to sell commercial type uranium and a signed agreement is expected to be in hand by July 1998.[11] While the total supply and demand picture is not known and credit may be somewhat overstated, failure to realize these savings would not in itself affect a decision on alternatives.

The question of whether the processing/co-disposal baseline cost estimates included handling the process waste stream as well as the product was raised during the Augusta meeting. The costs of handling, converting the waste to glass logs, and repository disposal are included in the estimates.[12]

Question 3. Are the cost and schedule estimates developed by DOE for the alternative processing options suitable for budget planning purposes?

Response. The estimates and schedules contained in the Task Force Report are not of budget quality. As discussed under question 2 above, the schedules appear to be unachievable even with universal DOE support and absence of controversy. That condition does not exist, because during the week of November 24 the budget disappeared and then was restored during the week of December 1.[13] Further, the cost estimates, while sufficient to pick among alternatives for further development, are not refined to the point to support budget expectations for line item projects.

As discussed under question 2, the Transfer Facility project was not submitted as a part of the FY98 DOE budget, and there has been no exposure of the project to the Congress to determine support there.[14] The program has been discussed with the Nuclear Regulatory Commission, but work on licensing of facilities has not yet begun. The December 1997 cost study indicates facilities will be constructed to USNRC standards but will not be licensed.[15]

In the time since the Task Force Report was published a pre-conceptual design of the Transfer Facility has been performed by Bechtel.[16] The roughly $240 million cost is about 10 percent under the like facility estimate in a non-proliferation cost study prepared in July 1997.[17] This bottoms-up estimate has the type of detail required to support a line item project and is now available for that purpose. In addition, studies on the modification and use of existing facilities at L Basin for the receipt portion of the Transfer Facility are being revisited.[18]

The cost estimates prepared to support the non-proliferation study show that while there have been adjustments as a result of better information and more detail available than when the original Task Force report was prepared, by far the greatest increase in life cycle costs among all alternatives is in operational costs. These costs increased by a factor greater than 2:1 and are responsible for the bulk of the cost growth.[19] While the unit costs for processing and other radioactive materials handling are well known, the budget questions will need to focus on the drivers of these costs (i.e., manpower and time). Again, the relative positions of alternatives remain unchanged.

Question 4. Has DOE considered the costs of program delays in its budget development or budget planning for this program?

Response. The costs contained in the Task Force report do not reflect program delays since they were prepared for a schedule with admittedly forced dates. At the Augusta meeting it was reported that the DOE decision on whether to proceed and with what alternatives is expected by October 1999,[20] and a project will be submitted either as an FY00 privatization project or an FY01 line item project.[21]

The cost estimates in the July 1997 and December 1997 non-proliferation cost studies have been adjusted for programming as currently foreseen. Inflation factors used appear to be reasonable.

Question 5. Are the cost and schedule estimates for implementing the alternative processing options consistent with DOE procedures and systems? If not, has DOE identified what changes must be made to achieve its cost and schedule targets?

Response. Cost estimate development and schedules contained in the Research Reactor Task Force Report are not consistent with DOE procedures because the forced schedule mentioned in questions 2, 3, and 4 did not permit development of budget quality estimates. The revised schedules provided at the Augusta meeting now support adherence to DOE procedures.

Estimates have been prepared for submission of the project under the privatization program.[22] There is no evidence of significant waivers of environmental, safety, and health procedures, DOE site work complications, and risk assumption, which would seem to be necessary to support the type of cost savings projected. The project was included in the FY99 privatization program but was dropped from the list in an FY99 pass-back. Five of seven projects submitted were reprogrammed as line item projects. DOE has yet to sign a privatization contract requiring debt financing, and experience to date with funded privatization contracts has been poor.[23] The December 1997 estimates include costs for privatization, including debt financing.[24] While factors are scaled to account for varying levels of relative risk, the track record to date doesn't answer the question of whether, in fact, investors will be willing to support such a venture. Pursuing this means of executing the project appears to guarantee additional delays (personal opinion).

Detailed scheduling of work for the processing canyons has been accomplished since publication of the Research Reactor Task Force Report. It has been confirmed that options involving reprocessing addressed in the report can be accommodated in the overall workload consistent with DOE procedures.[25] The December 1997 cost estimates also show a $240 million life cycle cost saving for the reprocess/co-disposal option.[26] This appears to be an attractive option, but the policy issue on whether to exercise it must be decided by DOE.

The Research Reactor Task Force Report recommends a project approach to the program.[27] If the line item project route is chosen, DOE procedures provide that critical milestone decisions for projects under $500 million are made by the local DOE site office.[28] Both a project approach and local decision making authority are essential to timely execution of the program (personal opinion).

Question 6. Are the cost and schedule milestones that are laid out in the Research Reactor Task Force Report for selecting and implementing an alternative processing option being met?

Response. The schedule milestones laid out in the Research Reactor Task Force Report are not being met for reasons discussed under questions 2 through 5 above.[29] Since the project has slipped and cost estimates have been revised, cost performance cannot be evaluated yet.

Revised schedules presented at the Augusta meeting appear to be achievable with strong DOE support and commitment (personal opinion).

A non-proliferation study currently under preparation is not expected to make go/no-go recommendations on any alternatives.[30] It should, however, assist in answering the political question on whether and how much reprocessing to do in order to support a timely decision on which alternatives to pursue and meet budget programming windows.

References

[1] Devine, J., et al., *Technical Strategy for the Treatment, Packaging, and Disposal of Aluminum-Based Spent Nuclear Fuel*, a report to the U.S. Department of Energy (DOE) by the Research Reactor Spent Nuclear Fuel Task Team, June 1996, Vol. II, pp. C-38,54,57. Also breakout session at Augusta, Georgia meeting, December 3, 1997 (John Hurd, WSRC/Bechtel).

[2] Ibid., Table C7.2-2f.

[3] Ibid., pp. C-124-127.

[4] Breakout session at Augusta, Georgia meeting, December 3, 1997 (John Hurd, WSRC/Bechtel).

[5] Breakout session at Augusta, Georgia meeting, December 3, 1997 (Randy Polnick, DOE-SR).

[6] Dupont, M.E., et al., *Savannah River Site FY97 Spent Nuclear Fuel Interim Management Plan (U)*, October 1996, p. 15.

[7] Briefing at Augusta, Georgia meeting, December 2, 1997 (Jack Devine, Polestar).

[8] Briefing at Augusta, Georgia meeting, December 2, 1997 (Joe Krupa, WSRC).

[9] Krupa, J.F., *Savannah River Site Aluminum-Clad Spent Nuclear Fuel Alternative Cost Study Rev 1 (U)*, December 12, 1997, p. 25.

[10] Devine, J., et al., *Technical Strategy for the Treatment, Packaging, and Disposal of Aluminum-Based Spent Nuclear Fuel*, a report to the U.S. Department of Energy (DOE) by the Research Reactor Spent Nuclear Fuel Task Team, June 1996, Vol. I, p. 58.

[11] Breakout session at Augusta, Georgia meeting, December 3, 1997 (John Dickinson, WSRC).

[12] Breakout session at Augusta, Georgia meeting, December 3, 1997 (Joe Krupa, WSRC).

[13] Briefing at Augusta, Georgia meeting, December 2, 1997 (Jon Wolfstal, DOE-HQ).

[14] Breakout session at Augusta, Georgia meeting, December 3, 1997 (John Hurd, WSRC/Bechtel).

[15] Krupa, J.F., *Savannah River Site Aluminum-Clad Spent Nuclear Fuel Alternative Cost Study Rev 1 (U)*, December 12, 1997, p. 25.

[16] Breakout session at Augusta, Georgia meeting, December 3, 1997 (Jane Carter, WSRC/Bechtel).

[17] Briefing at Augusta, Georgia meeting, December 2, 1997 (Joe Krupa, WSRC).

[18] Breakout session at Augusta, Georgia meeting, December 3, 1997 (Mark Barlow, WSRC).

[19] Devine, J., et al., *Technical Strategy for the Treatment, Packaging, and Disposal of Aluminum-Based Spent Nuclear Fuel*, a report to the U.S. Department of Energy (DOE) by the Research Reactor Spent Nuclear Fuel Task Team, June 1996, Vol. II, Table C8.2f. Also Krupa, J.F., *Savannah River Site Aluminum-Clad Spent Nuclear Fuel Alternative Cost Study Rev 1 (U)*, December 12, 1997, Table D-1f.

[20] Briefing at Augusta, Georgia meeting, December 2, 1997 (Karl Waltzer, DOE-SR).

[21] Breakout session at Augusta, Georgia meeting, December 3, 1997 (Randy Polnick, DOE-SR).

[22] Dupont, M.E., et al., *Savannah River Site FY97 Spent Nuclear Fuel Interim Management Plan (U)*, October 1996, pp. 4, 27f.

[23] Breakout session at Augusta, Georgia meeting, December 3, 1997 (Randy Polnick, DOE-SR).

[24] Krupa, J.F., *Savannah River Site Aluminum-Clad Spent Nuclear Fuel Alternative Cost Study Rev 1 (U)*, December 12, 1997, p. 21f.

[25] Breakout session at Augusta, Georgia meeting, December 3, 1997 (John Dickinson, WSRC).

[26] Krupa, J.F., *Savannah River Site Aluminum-Clad Spent Nuclear Fuel Alternative Cost Study Rev 1 (U)*, December 12, 1997, p. 25.

[27] Devine, J., et al., *Technical Strategy for the Treatment, Packaging, and Disposal of Aluminum-Based Spent Nuclear Fuel*, a report to the U.S.

Department of Energy (DOE) by the Research Reactor Spent Nuclear Fuel Task Team, June 1996, Vol. I, p. 58.

[28] Breakout session at Augusta, Georgia meeting, December 3, 1997 (Randy Polnick, DOE-SR).

[29] Briefing at Augusta, Georgia meeting, December 2, 1997 (Karl Waltzer, DOE-SR).

[30] Briefing at Augusta, Georgia meeting, December 2, 1997 (Jon Wolfstal, DOE-HQ).

Topic: Processing and Remote Systems
Consultant: Harry Harmon

Introduction

We were asked to participate in the presentations on December 2-3, 1997, and study the materials provided in order to address a set of specific questions that were given to us. Joe Byrd and I served as a subteam on Processing/Remote Handling. In addition to the reports and handouts received on the first day of the meeting, we met with a number of technical personnel involved in the program on the second day. These personnel included program participants from Savannah River Site, Argonne, Idaho National Engineering and Environmental Laboratory, and interested members of the public. My response to the questions is provided below. Mr. Byrd will provide more detail than I have offered in the area of remote handling.

Responses

Q: For each processing option evaluated by DOE, are the processing steps used as the basis for assessment and comparison (other than direct co-disposal) technically credible? That is, are they likely to work as described and produce the products and results assumed?

A: I believe that enough is known about each process option to conclude that they are all technically credible; i.e., the physical and chemical processes employed should work in principle. However, whether they will work as described in a remote plant environment is more difficult to state with certainty. The latter part of this question depends more on the level of development and prior experience with the process steps. Unit operations such as packaging, mechanical size reduction, melting, dissolution, vitrification, and electrorefining of uranium and aluminum are well known and demonstrated.[1,2] The electrometallurgical process was depicted as having little or no secondary waste, but similar processes at Rocky Flats have generated significant quantities of salt waste and other residues.[3] A key development need for the electrometallurgical process is actual demonstration of the ability to recycle all potential waste streams. I believe that the Glass Material Oxidation and Dissolution System (GMODS) and Plasma Arc Treatment will require extensive development programs to support a reliable plant-scale process. For GMODS, I believe that feeding pieces of fuel elements to the melter, pouring the glass product, and off-gas processing will be challenging steps. In the case of plasma arc treatment, feeding fuel elements, remote maintenance of the rotating furnace, control of the ceramic waste form composition, and off-gas processing will be significant development concerns.[1]

Q: Are there other processing options that should be considered by DOE for disposition of aluminum-based spent fuel?

A: Reprocessing of portions of the spent fuel in 221-H canyon at the Savannah River site should be given more consideration in this study. While the timing of fuel receipts may preclude processing all the fuel in 221-H (even if policy considerations allowed it), the program could be simplified by reprocessing portions of the fuel. First, part of the fuel could be reprocessed to alleviate basin capacity concerns. The resulting purified HEU uranium solution could be diluted to the desired level (less than 20 percent for proliferation concerns or less than 5 percent for LWR fuel use) and fission products would go to the Defense Waste Processing Facility (DWPF) and, eventually, to the repository. How much is processed would depend on shipping schedules, basin space, and demand for LEU uranium for LWR fuel. Secondly, the program would be simplified by eliminating small quantities of U-Al fuels that are significantly different in size from conventional Materials and Test Reactor (MTR) fuel elements.[1] For example, long rods (like NRU/NRX fuel) will complicate design of fuel handling and feed preparation steps and will require some size reduction even for direct co-disposal. (Some believe that non-standard fuel dimensions are a greater problem for direct co-disposal than for the processing options.) All these U-Al fuels, uranium metal in aluminum cans, and UO_2 in aluminum cans are chemically compatible with SRS canyon processes (although the UO_2 powder materials will require some special nuclear safety controls during dissolution). Thus, they could be eliminated from design considerations by reprocessing them.

Two variations briefly considered in the Technical Strategy Report[1] should, in fact, be given further evaluation: (1) Small casks designed for interim dry storage and transportation could be used by the reactor sites for shipment of MTR fuel to SRS. Casks, such as the GNB CASTOR MTR cask, are commercially available and should be considered in cost analyses of storage options. (2) Not shipping INEEL SNF to SRS reduces SRS storage needs and minimizes unnecessary transportation. If direct co-disposal is selected, INEEL will have the packaging facilities required, based on its role with stainless steel and zircalloy-clad fuels.[4]

Q: Do inner container designs appear adequate to contain the waste forms resulting from the various processing options?

A: Although I did not examine this in great depth, I see no reason why the designs would not be adequate. Clearly, the processing options must size their product containers appropriately, but this is straightforward.

Q: Are they overdesigned for the intended application?

A: No. The canisters were described as being fabricated of steel with neutron poison inserts as required. DWPF canisters are stainless steel, so it seems appropriate that the SNF canister should be of similar durability.

Q: Are DOE's basic material handling plans, pool use, and other facility needs reasonable, and are remote handling technologies available to meet these needs?

A: The transfer and storage facility concepts[1] employed standard equipment and conventional techniques for remote material handling. For direct co-disposal, remote size reduction equipment will be required to accommodate the dimensional restrictions of the inner containers for other than standard MTR fuel elements. All required remote handling technologies are readily available, and advancements in this field continue to be made.

Q: Are the technical requirements for the alternative processing options sufficiently well defined so that reasonable judgments can be made about the likelihood of success of implementing them?

A: The processing options can be divided into three families: (1) HEU dilution technologies; (2) advanced treatment technologies; and (3) options involving canyon processing. Technical requirements for each family of options have been defined, at least in general terms. The Technical Strategy[1] identified four criteria that all process options must meet to be considered:

• Development work must be completed by 2000 (this is no longer viewed as required by DOE).
• Funding required during the development period (e.g., during the first five years) must be within that reasonably expected to be available in that time frame.
• The waste form must be compatible with anticipated repository requirements.
• The treatment technology cannot present any environmental, safety, and health operational concerns.

Also, requirements for repository disposal of the Al-SNF form are listed in Reference 2, but most of those criteria relate to the canister and waste package.

However, I was not able to find a complete set of process requirements or product requirements that all options must meet to be successful. (Such a document may exist, but I have not seen it.) For example, if all SNF HEU must be diluted to less than 20 percent before going to the repository, then direct disposal and direct co-disposal are eliminated. Also, if separation of fission products from fissile material is forbidden, then all canyon processing options and electrometallurgical treatment are eliminated. Without firm requirements identified that all processes and products must meet, the options can only be evaluated as possible approaches, each with potentially different end products.

Also, since product requirements are not specifically identified, evaluations of characterization needs must be very broad and comprehensive at this juncture.[5] Extensive dialogue and teamwork between the spent fuel program and the repository will be required to develop an achievable and acceptable characterization plan. I would recommend use of commercially available instrumentation for burnup and fissile content measurements on fuel elements. With these data, fission product content can be calculated with sufficient accuracy to provide characterization via process knowledge.

Q: Are there large differences in likelihood of success of implementing the various process options?

A: Yes. To be considered successfully implemented, the process option must be capable of being implemented in an operating facility within the budget and schedule constraints. (Given sufficient time and funding, all options could be implemented in my opinion.) Thus, the likelihood of success in this context is inversely related to the extent of technology development needed. Based on my judgment and the information in the technical strategy document,[1] I would rank them as follows (highest likelihood of success at top of list):

1.   Processing/co-disposal
2.   Direct disposal and direct co-disposal
3.   Press/dilute and melt/dilute
4.   Dissolve and vitrify
5.   Electrometallurgical
6.   Plasma arc
7.   GMODS

There are not large differences in likelihood of success between the first three groups, but the last four will require significantly more effort to be successful.

References

1. "Technical Strategy for the Treatment, Packaging, and Disposal of Aluminum-Based Spent Nuclear Fuel," Volumes I and II, June 1996.
2. L. Sindelar et al. "Alternative Aluminum Spent Nuclear Fuel Treatment Technology Development Status Report (U)," WSRC-TR-97-00345, October 1997.
3. Implementation Plan for Defense Nuclear Facility Safety Board Recommendation 94-1. May 1994.
4. P. Hoskins et al. "Technical Strategy for the Management of INEEL Spent Nuclear Fuel," March 1997.
5. E. Skidmore et al. "Task Plan for Characterization of DOE Aluminum Spent Nuclear Fuel (U)," SRT-MTS-97-2004, January 31, 1997.

Topic: Nuclear Criticality Safety
Consultant: Valerie L. Putman

Introduction

Aluminum-based Spent Nuclear Fuel (Al-SNF) is to be collected at the Savannah River Site (SRS) for treatment and interim storage, then shipped to a permanent repository for final storage. Currently stored at SRS, various research reactors and U.S. Department of Energy (DOE) sites throughout the United States, and various foreign research reactors, Al-SNF is in a wide variety of configurations. Most Al-SNF is highly enriched in ^{235}U at beginning and end of life. Al-SNF is therefore considered to be a greater criticality safety concern throughout treatment, interim storage, transport, and final storage than Spent Nuclear Fuel (SNF) from commercial power plants.

SRS management must adequately identify, characterize, and weigh options for DOE to select a path forward for disposing this fuel. SRS staff must work closely with organizations that currently have the Al-SNF and with repository personnel to ensure that options adequately address all issues for the Al-SNF itself and for the Al-SNF in the repository in an efficient, cost-effective manner.

SRS adequacy in addressing criticality safety issues of the options is reviewed here. Two criticality safety questions were specified for this review (Levenson and Crowley, 1997):

Question 1: What are the significant criticality issues that must be considered during processing, interim storage of the waste form after processing, and shipment of the waste form to a repository? Has DOE adequately addressed these issues in its technology planning?

Question 2: Do any of the waste forms produced by the alternative processing options pose significant internal or external criticality hazards in a repository—either from material degradation in the waste container or in the near field of the repository after the container is breached—relative to commercial spent fuel or vitrified high-level waste? NOTE: comments on the use of poisons or isotopic dilution are appropriate as are comments on filling the void space in the canister so as to limit the volume of water that could be present in case of canister leakage.

Conclusion

Adequate criticality safety can be assured for configurations and activities to treat, move, and store DOE aluminum-based spent nuclear fuel (Al-SNF). Criticality safety analysis methodologies are well developed, and existing computer codes and neutron cross section libraries appear sufficient. Although few, if any, specific criticality safety evaluations are complete, scoping work indicates sufficient safety can be provided by standard means such as limiting fissile quantity, including neutron absorbers, and for the near-term, limiting configurations. In addition, information from in-progress and planned tests will be used to determine if additional criticality safety resources are needed.

SRS presenters identified a co-disposal option as currently most favored. Strategies are identified to prevent a critical excursion for this option during all stages of treatment, interim storage, and final disposal. Controls include fixed neutron absorbers in canister baskets for some Al-SNFs. However, continued mixing of fissile material and neutron absorber is less certain after many millennia, when fuel and canisters are fully degraded and material might migrate outside the repository. Although information to date indicates the consequences of a critical excursion with Al-SNF at this point would be negligible, accident prevention apparently is still considered very important.

Therefore, this reviewer believes, if critical excursion prevention continues to be a very high priority after fuel and canisters are fully degraded and material might migrate outside the repository, Al-SNF treatment(s) that significantly dilute ^{235}U and/or introduce significant neutron absorbers in the fuel matrix should be selected.

Discussion

Except for scoping calculations, little fuel-specific or activity-specific criticality safety work is complete for proposed activities with these Al-SNFs. More specific work is premature until fuel-treatment and storage-configuration options are narrowed further. It is sufficient to ensure that adequate methodology (codes and cross-section libraries; tool selection methods; modeling, calculation, validation, and documentation requirements and methods; criticality accident scenario identification; and criticality safety contingency analysis) and information (fuel, facility, and treatment descriptions) will be available to perform specific criticality safety evaluation(s) when needed. Current availability is not required if measures are taken to identify and obtain necessary items in a timely manner to support option selection and specific evaluations.

Criticality Safety at Savannah River Site (SRS)

Many proposed activities with Al-SNF at SRS are similar to past activities, there and elsewhere, with aluminum-based nuclear fuels. These activities include underwater interim SNF storage for some Al-SNF types, packaging or repackaging into fuel-specific baskets in approximately 17-inch-diameter Al-SNF canisters, drying Al-SNF and/or loaded canisters as needed, normally dry Al-SNF storage, and associated fissile material handling and transportation. For criticality safety purposes, most of these activities should represent minor perturbations from previously evaluated conditions. Although, these activities must be evaluated to develop specific criticality safety limits and critically safe designs, or to show applicability of limits and designs established for Al-SNF already at SRS, SRS criticality safety methodology should be adequate for these evaluations.

Additional treatment options discussed are dissolve and vitrify (glass or ceramic waste), electro-metallurgical uranium separation, glass material oxidation and dissolution, melt and dilute (metal waste), plasma-arc vitrification (ceramic waste), press and dilute (metal waste), and, as a baseline, continued wet-chemical uranium separation. With the exception of the press-and-dilute option, these treatments present opportunities to dilute ^{235}U and/or introduce neutron absorbers in the fuel matrix itself, which would better assure criticality safety in the final repository after many millennia when the fuel matrix and canisters are completely degraded. Options involving uranium separation also include uranium reuse, if appropriate, based on economics and non-proliferation policies.

With the exception of wet-chemical processing and, possibly, dissolve and vitrify options, some steps of these treatments could represent major perturbations from previously evaluated conditions. However, U.S.-wide experience in developing new fissile material processes indicate that adequate criticality safety control should be possible through vessel and/or container geometry, fissile mass or dilution limits, fixed or soluble neutron absorbers, and/or moderation limits for each step of these further treatment processes. The problem is usually one of adequately balancing criticality safety with process efficiency during process development.

Criticality safety methodology at SRS is well developed due to its many years of nuclear experience with reactor operations, fuel storage, fuel processing via chemical reactions, and associated fissile material handling. Its past missions include significant experience with aluminum-based fuels. Therefore, at least some of criticality safety staff have considerable experience with software, cross-section data, and fuel-characterization data available for evaluating such fuels (see Gough et al., 1997, as an example).

Criticality-safety-evaluation methodology at SRS includes a well-developed validation program which should identify and adequately compensate for any problems that might exist in codes and/or cross-section data. Staff experience and the validation program itself appear sufficient to ensure calculation validation effectiveness (Kimball and Trumble, 1997; Chandler and Trumble, 1997).

None of the discussed Al-SNF activities at SRS are anticipated to involve conditions under which basic nuclear data and/or data processing are currently questioned (for example, ^{235}U very highly diluted with aluminum, ^{235}U with massive aluminum reflectors, or uranium in the resonance neutron energy range). Current SRS validation practices should be sufficient to identify if a less-than-desirable code and cross-section combination is used for particular conditions (for example, ^{235}U in a fast energy system with SCALE 4.3 and ENDF/B-VI.3 cross-sections). Therefore, available information and data should be sufficient to perform and validate the necessary criticality safety evaluations. Additionally, ongoing tests should provide sufficient information to identify conditions that significantly deviate from expected conditions, enabling staff to determine if criticality safety tools are adequate in a timely manner.

Criticality Safety on Public Roads, Transport to Final Repository

Criticality safety requirements for transportation of fissile material over public roads are well established and are updated as needed. Transportation requirements specifically identify many conditions that must be evaluated, relying less upon the evaluating or operating organization to identify all important conditions.

Neither SRS nor Yucca Mountain criticality safety personnel typically perform many transportation evaluations. However, cask vendors, license holders, and several DOE contractor sites have well established criticality safety methodology to satisfy transportation criticality safety evaluation requirements. The methodologies are similar in many respects to the SRS criticality safety methodology but are specifically tailored to the types of fissile materials with which these other organization are concerned. General principles apply but some specifics (for example, detailed modeling methods) might not apply.

Whichever organization performs transportation analyses for treated Al-SNF, its methodology will need to be reviewed for applicability to the treated fuel. Few if any problems are expected in developing a methodology because most options would produce a treated Al-SNF less reactive than previously shipped configurations of respective beginning of life aluminum-based fuel and/or untreated Al-SNF. Although there are larger-than-typical uncertainties in reflection cross sections of lead and iron (materials in transportation casks that would not necessarily be addressed elsewhere), established transportation criticality safety methodologies must already have resolved any problems caused by these uncertainties.

A major difference between SRS and typical transportation criticality safety methodology is procedures for determining calculation-method bias and uncertainty. Although SRS's procedure is less conservative than typically used for transportation evaluations, the SRS procedure is defensible and valid.

Criticality Safety at Final Repository

Al-SNF treatment and packaging options to be implemented at SRS greatly affect strategies to assure criticality safety for specific fuel and storage configurations in the final repository, and vice-versa. Candidate options result in fuel matrices and configurations ranging from packages enveloped by commercial-SNF packages to pockets of highly enriched uranium surrounded by highly radioactive waste, depleted uranium, and/or commercial SNF. In the latter case, SRS prefers a co-disposal option because it limits pocket size, providing a more assured fissile material dilution than direct disposal as canisters degrade over millennia.

Repository criticality safety methodology is developed for final repository at the Yucca Mountain site. Initially developed for commercial SNF, it was recently revised to address highly enriched DOE SNF, including Al-SNF. Revisions included expanding the validation database and developing strategies for handling conditions not already addressed by methodology for commercial SNF (Thomas et al., 1997).

Repository criticality safety work to date for Al-SNF primarily evaluates a co-disposal option. Of SRS favored options, this one is judged to be most vulnerable to a criticality accident because the waste form is repackaged fuel assemblies, still qualifying as highly enriched uranium in which each assembly, if flooded, is nearly optimally moderated. If adequate criticality safety can be assured for this option, adequate safety can be assured, perhaps with different strategies, for other favored but less vulnerable options.

These Al-SNF criticality safety studies are based on two SRS-identified representative Al-SNFs, Massachusetts Institute of Technology (MIT) and Oak Ridge Research (ORR) reactor fuel assemblies (Doering and Gottlieb, 1997). Some Al-SNFs are more reactive than the representative fuels (for example, University of Missouri Research Reactor (MURR) fuel) but might not be adequately representative of other Al-SNF characteristics (Sentieri, 1996, p. 69). Unless additional evaluations are performed for more reactive fuels, it will probably be necessary to limit canister loadings to ensure each fuel array in a canister is no more reactive than the most reactive loading of representative Al-SNF. Such a limitation is not necessarily inefficient depending on relative individual and cumulative fuel assembly volumes.

With the exception of specifically evaluating the most reactive Al-SNF fuel assemblies, evaluated conditions were selected to envelope a wide variety of Al-SNF assemblies and repository conditions. This strategy is adequate and economical for DOE to determine a path forward and might be adequate for final evaluations if the co-disposal option is selected.

A comparison of evaluated critical-experiments and repository-condition characteristics, with the representative fuels, indicates there are sufficient experiment data to validate most enveloping repository conditions (Anderson, 1977; Doering and Gottlieb, 1997; Gottlieb, 1997; Gottlieb et al., 1997). In some cases, calculations might be more conservative than absolutely necessary. Additional experiment data are not essential but such data might allow minor storage efficiency improvements.

There is a general lack of critical experimental data for extremely dilute ^{235}U systems and for fissile systems in the intermediate neutron energy range. This is not currently a concern because calculations indicate repository conditions involving these characteristics have extremely low subcritical k_{eff}s (Gottlieb, 1997; Gottlieb et al., 1997). In these cases, calculation validation is much less of a concern. For example, if a condition's calculated k_{eff} is 0.3, it is more important to demonstrate that the condition's actual k_{eff} satisfies the required margin, in this case does not exceed 0.95, than to demonstrate that the calculated k_{eff} is within a few percent of the actual k_{eff}.

Intact Canisters and Al-SNF, Initial Repository Conditions

Traditional criticality safety requirements, concerns, and issues apply to the final repository initially because workers could be at risk if a criticality accident were to occur and because evaluations and corrective actions could be undertaken in a reasonable manner if problems develop. Initial co-disposal conditions involve handling and storing intact or nearly intact (non-leaking, very close to initial configuration) Al-SNF canisters, each surrounded by five co-disposed waste canisters.

Specific canister-array configurations probably differ but methodologies and data used in addressing criticality safety for treated

Al-SNF in interim storage and for transportation should apply to evaluations for initial repository conditions. Although possibly less well known, neutron reflection and moderation properties of materials that could credibly be between and/or around Al-SNF canisters in the repository should be no worse than materials (water, concrete, transportation cask, and adjacent fuels) that must be considered for these earlier activities.

Initial (Phase 1) criticality safety analysis of intact or nearly intact Al-SNF canisters and fuel is nearly complete for the co-disposal option with the two representative fuels (Doering and Gottlieb, 1997; Gottlieb, 1997). One representative fuel, ORR, is sufficiently reactive to require fixed neutron absorbers for a critically safe, efficient Al-SNF canister loading under flooded conditions (Doering and Gottlieb, 1997). If ORR fuel with fixed neutron absorbers is acceptable, it should be possible to qualify canisters of the more reactive Al-SNFs, possibly with fewer fuel assemblies and the same neutron absorber or with the same number of fuel assemblies and more effective neutron absorbers (for example, thicker metal plates or higher concentrations in metal).

Degraded Canisters and/or Al-SNF, Long Term Repository Conditions

Long term repository criticality safety is handled in a non-traditional manner because there would eventually be no mechanism to detect developing problems, and because the location is very well shielded if an accident were to occur. In this case, evaluations examine environmental consequences more closely than human radiological exposure because humans are not at direct risk from a criticality accident. An inadvertent critical excursion is still undesirable and preventative measures are required. However, an extremely low-probability event might be acceptable if environmental consequences are negligible.

Initial criticality safety analysis of degraded Al-SNF canisters and/or Al-SNF fuel (Phase 2) is nearing completion for the co-disposal option with two representative fuels (Gottlieb et al., 1997). Controls to minimize critical excursions in or near a canister are considered in this phase. In some cases it is important for criticality safety to ensure that, where neutron absorbers were required for intact fuel, absorbers continue to be "mixed in" with degrading fuel. Gadolinium in carbon steel baskets is the currently preferred absorber because it is expected to degrade in a manner that should encourage mixing with degraded fuel.

Criticality safety analysis is initiated for storage conditions after millennia; canisters, baskets, and fuel are theorized to be fully degraded and completely uncontained (Phase 3). It is again preferred that, where neutron absorbers were required for intact fuel, absorbers continue to be mixed with fissile material. However, mixing is less assured at this stage. This reviewer believes that, if critical excursion prevention continues to be a very high priority for this phase, Al-SNF treatment(s) that significantly dilute ^{235}U and/or which introduce significant neutron absorbers in the fuel matrix should be selected to better assure continued mixing.

Criticality accident prevention is desirable but prevention methods are less assured. Critical excursion consequences therefore are more important for ensuring acceptably low risk during Phase 3. Information to date indicates that, if a critical excursion were to occur only in the highly enriched SNF, humans would still be very well shielded from the excursion. Additionally, resultant increases in fission products would be negligible compared to the already large inventory from commercial SNF, even when considering decay of the commercial SNF fission product inventory before this hypothetical excursion.

Summary

Adequate criticality safety can be assured for configurations and activities to treat, move, and store DOE Al-SNF. Criticality safety analyses methodologies are well developed and existing computer codes and neutron cross section libraries appear sufficient. Scoping work indicates sufficient safety can be provided by standard means such as limiting fissile quantity, including neutron absorbers, and, for the near-term, limiting configurations. Information from in-progress and planned tests will be used to determine if additional criticality safety resources are needed.

Most criticality safety issues for proposed activities to treat, store in interim facilities, and ship Al-SNF will be the same as, or very similar to, criticality safety issues already addressed for existing SRS fuel processing, storage, transfer, and transport activities. Methods for addressing criticality safety issues under such circumstances and for such processes are well established and apparently adequate (specifically, acceptable to the operating contractor and to DOE). Although analyses to define specific criticality safety limits are not yet initiated, proposed activities should not pose a technical challenge to this analysis methodology. Also, although there is a difference between SRS and typical transportation criticality safety procedures for determining bias, the SRS procedure is defensible and valid.

Activity similarity exceptions involve advanced treatment options because some aspects of these treatments might challenge adequacy of existing SRS data and/or expertise. However, SRS criticality safety analysis methodology would still apply, but updated, upgraded, or different specific tools (modeling conventions, codes, cross-section libraries) might be needed. Planned and in-progress testing should soon identify Al-SNF characteristics for process stages of concern, at which point any need for updated, upgraded, or different criticality safety tools can be determined. Information to date indicates current tools are sufficient.

In most cases, treatment and packaging options result in an Al-SNF waste that is significantly more reactive and has a significantly different fuel composition than commercial SNF waste. At this time, criticality safety analyses of Al-SNF in a permanent repository focus on a co-disposal option. Of SRS favored options, this one will result in the most reactive permanent storage configuration. If adequate criticality safety can be assured for this configuration and waste form, then staff should be able to assure adequate criticality safety for less reactive configurations and waste forms.

Appropriate criticality safety methodology including validation is developed for the Yucca Mountain final repository. Preliminary evaluations are nearly complete, based on two representative Al-SNFs, the selected co-disposal option, and a few conditions selected to envelope all credible conditions identified for Al-SNF in the repository. A comparison of characteristics for these enveloping conditions and for critical experiments in the repository's validation database indicates the methodology is adequate.

Although representative Al-SNFs do not include the most reactive Al-SNF assemblies, preliminary evaluations are sufficient to indicate adequate criticality safety can be established for this option. Depending on specific baskets and storage configurations for the more reactive fuels, these preliminary evaluations might qualify as enveloping evaluations if the co-disposal option is selected.

Final repository critical excursion prevention relies in some cases on neutron absorbers, which should be mixed with the Al-SNF, either between assemblies, between assembly components, or within the fuel matrix itself. After many millennia, fuel and baskets could be completely degraded and materials would not be contained. Under such conditions, continued mixing of neutron absorbers and ^{235}U is less assured. Information to date indicates that the consequences of a critical excursion would be negligible and should be acceptable. However, if critical excursion prevention is paramount even at this fully degraded stage, this reviewer recommends treatment option(s) that dilute ^{235}U and/or introduce neutron absorbers in the fuel matrix be selected to better assure continued fuel and absorber mixing.

References

Anderson, M. J. 1977. "Summary Report of Laboratory Critical Experiment Analyses Performed for the Disposal Criticality Analysis Methodology," B00000000-01717-5705-00076, Rev. 0. Civilian Radioactive Waste Management System Management & Operating Contractor: Las Vegas, Nevada (September 4, 1997).

Chandler, John R., and E. Fitz Trumble. 1997. "Use of Bias, Uncertainty, and Subcritical Margins at the Savannah River Site." In Proceedings, Criticality Safety Challenges in the Next Decade, Chelan, Washington, September 7-11, 1997. American Nuclear Society: La Grange, Illinois (September 1997), pp. 262-267.

Doering, Thomas W., and Peter Gottlieb. 1997. "Evaluation of Codisposal Viability for Aluminum-Clad DOE-Owned Spent Fuel: Phase 1, Intact Codisposal Canister," BBA000000-01717-5705-00011, Rev. 1. Civilian Radioactive Waste Management System Management & Operating Contractor: Las Vegas, Nevada (August 15, 1997).

Gottlieb, P[eter]. 1997. "Degraded Waste Package Criticality: Summary Report of Evaluations through 1996," BBA000000-01717-5705-00012, Rev. 0. Civilian Radioactive Waste Management System Management & Operating Contractor: Las Vegas, Nevada (August 28, 1997).

Gottlieb, P[eter], et al. 1997. "Waste Package Probabilistic Criticality Analysis: Summary Report of Evaluations in 1997," BBA000000-01717-5705-00015, Rev. 0. Civilian Radioactive Waste Management System Management & Operating Contractor: Las Vegas, Nevada (September 16, 1997).

Gough, Sean T., et al., 1997 "Process Error-Induced Criticality Accident Analysis for Savannah River Site Receiving Basin for Offsite Fuel." In Proceedings, Criticality Safety Challenges in the Next Decade, Chelan, Washington, September 7-11, 1997. American Nuclear Society: La Grange, Illinois (September 1997), pp. 306-309. (Primarily used as an example of SRS criticality safety methodology.)

Kimball, Kevin D., and E. F[itz] Trumble. 1997. "Statistical Methods for Accurately Determining Criticality Code Bias." In Proceedings, Criticality Safety Challenges in the Next Decade, Chelan, Washington, September 7-11, 1997. American Nuclear Society: La Grange, Illinois (September 1997), pp. 247-254.

Levenson, Milt, and Kevin Crowley. 1997. "Instruction to the Technical Experts, Draft." Memoranda for Technical Options for Disposition of Al-Based Spent Nuclear Fuel. Board on Radioactive Waste Management, National Research Council: Washington D.C. (November 1997).

Sentieri, P[aul] J. 1996. "Criticality Safety Evaluation for Various Fuels Proposed for Dry Canning," LITCO Internal Report INEL-95/306. Lockheed Idaho Technologies Company: Idaho Falls, Idaho (April 1996).

Thomas, D. A., et al. 1997. "Disposal Criticality Analysis Methodology Technical Report," B00000000-01717-5705-00020, Rev. 1. Civilian Radioactive Waste Management System Management & Operating Contractor: Las Vegas, Nevada (September 4, 1997).

Other Documents Reviewed

ANS-8.1 Working Group. 1988. Nuclear Criticality Safety in Operations with Fissile Materials Outside Reactors, ANSI/ANS-8.1-1983(R1988). American Nuclear Society: La Grange Park, Illinois (approved October 7, 1983).

Bradley, Terry L., et al. 1996. "Direct and Co-Disposal Treatment Technologies," Appendix D. In Technical Strategy for the Treatment, Packaging, and Disposal of Aluminum-Based Spent Nuclear Fuel, Vol. 2. Research Reactor Spent Nuclear Fuel Task Team (June 1996).

Fisher, L. E., et al. 1988. Package Review Guide for Reviewing Safety Analysis Reports for Packaging, UCID-21218, Rev. 1. Lawrence Livermore National Laboratory: Livermore, California (October 1988), pp. 8-6–8-10.

Huria, H., and M. Ouisloumen. 1997. "ENDF/B-VI: ^{238}U Resonance Integral Reduction–A Closer Look." In Transactions of the American Nuclear Society, Vol. 76, American Nuclear Society: La Grange, Illinois (June 1997), pp. 329-330.

Lovett, Phyllis A., et al. 1996. "Criticality Control Bases for Repository Licensed Under 10 CFR Part 60," Appendix B. In Technical Strategy for the Treatment, Packaging, and Disposal of Aluminum-Based Spent Nuclear Fuel, Vol. 2. Research Reactor Spent Nuclear Fuel Task Team (June 1996). Prepared for U.S. Department of Energy, Office of Spent Fuel Management: Washington, D.C.

Savannah River Technology Center. 1997. "Alternative Aluminum Spent Nuclear Fuel Treatment Technology Development Status Report (U)," WSRC-TR-97-00345 (U). Westinghouse Savannah River Company: Aiken, South Carolina (October 1997).

Spencer, Robert R., et al. 1997. "Neutron Total and Capture Cross-Section Measurements on Aluminum at ORELA." In Transactions of the American Nuclear Society, Vol. 77. American Nuclear Society: La Grange, Illinois (November 1997), pp. 237-238.

TRW. 1995. "Preliminary Requirements for the Disposition of DOE Spent Nuclear Fuel in a Deep Geologic Repository," A00000000-00811-1708-00006, Rev 0. TRW Environmental Safety Systems Inc.: Vienna, Virginia (December 15, 1995).

TRW. 1997. "OCRWM Data Needs for DOE Spent Nuclear Fuel," A00000000-01717-2200-00090, Rev 2 Draft. TRW Environmental Safety Systems Inc.: Vienna, Virginia (September 1997).

U.S. Department of Energy. 1995. Facility Safety, DOE O 420.1 with Changes 1 and 2. U.S. Department of Energy: Washington D.C. (October 13, 1995; November 16, 1995; October 24, 1996), part 4.3.

U.S. Nuclear Regulatory Commission. Disposal of High-Level Radioactive Waste in Geologic Repositories, Part 60 in Title 10 of Code of Federal Regulations (10 CFR 60).

U.S. Nuclear Regulatory Commission. Packaging and Transportation of Radioactive Material, Part 71 in Title 10 of Code of Federal Regulations (10 CFR 71).

Weinman, J. P. 1997. "Monte Carlo Cross-Section Testing for Thermal and Intermediate $^{235}U/^{238}U$ Critical Assemblies–ENDF/B-V Versus ENDF/B-VI." In Transactions of the American Nuclear Society, Vol. 76. American Nuclear Society: La Grange, Illinois (June 1997), pp. 325-327.

Wright, R. Q. and L[uiz] C. Leal. 1997. "Benchmark Testing and Status of ENDF/B-VI Release 3 Evaluations." In Transactions of the American Nuclear Society, Vol. 77. American Nuclear Society: La Grange, Illinois (November 1997), pp. 232-234.

Topic: Proliferation Aspects of the Treatment Options
Consultant: David Rossin

Several points are made below based on my personal knowledge and experience over 43 years in nuclear reactor technology, materials and fuel cycle work. Much of that experience has involved licensed commercial nuclear power plants and fuel cycle facilities. Where the text reflects my personal opinions, it is so noted.

The discussion is divided into two main topics: (1) proliferation resistance and (2) assessment of alternative technologies under the Environmental Impact Statement process.

Proliferation Assessment

United States is a Weapons State—Safeguards Capability in the United States is internationally accepted. The Task Team Report included proliferation resistance for 5 percent of its Kepner-Tregoe evaluation. No significant differences between alternatives were identified. The report shows that further consideration of proliferation resistance is not needed for a meaningful comparison of alternatives. In my opinion, this is a reasonable approach. Because proliferation resistance is definitely a topic of concern to DOE, it would not be fitting to give the topic zero weight or to omit it, since that would not provide transparency for the analysis, and since the topic must be covered in the Programmatic EIS.

Proliferation Potential for Disposition Options. Obviously, all operations at SRL are conducted under DOE safeguards. Thus there is no proliferation risk associated with any of the actual operations that are under consideration.

All future operations including storage and handling of waste packages prior to insertion in the repository will also be conducted at safeguarded DOE sites. All transportation will also have to have appropriate safeguards. It is certainly reasonable at this time to assume that appropriate safeguards will be required regardless of the extent to which specific regulations or commitments exist at this time. In my opinion, there is confidence that appropriate regulations will be developed and applied when needed in the future, and therefore there should be no delays in the decision process because of transportation or handling safeguards.

Dilution of HEU may be needed for criticality considerations, but not for proliferation resistance at SRL. In my opinion, dilution of unique and costly HEU should not be done unless a valid and cost-effective case can be made for it. Physical security, accounting and safeguards protocols for handling HEU have been applied for decades, so a demonstration of diluting a quantity of HEU has no significance in international understanding of nonproliferation policy.

Identification of Separated Fissile Material as a Proliferation Risk

If DOE were to designate any separated fissile material as a proliferation risk, this could have extensive and costly ramifications. It could rule out separation of uranium rich in U-235 into a waste form suitable for disposal or require dilution at a point in the process that might not be necessary or cost-effective.

Such a designation could even preclude selection of co-disposal as an alternative. This does not appear to be a desirable result, based on the Task Team analysis. It might even force the choice to the melt-dilute alternative, even if other considerations point elsewhere.

In theory, it would be fine if all fissionable material were in forms or storage that meet either the "Spent Fuel Standard" or the "Stored Weapons Standard" described by the NAS report *Management and Disposition of Excess Weapons Material*. I believe there are commonsense levels on either side of each of these concepts that provide adequate safeguards for certain categories of fissionable materials. Neither of these concepts are really standards, in that they are not promulgated by any international or even national standards organization. They are valuable concepts since they represent known states, but are not to be applied blindly.

Container

The proposed co-disposal container has a canister surrounded by vitrified HLW. This container design has enough radioactivity to meet a self-protection criterion. However, as with the spent fuel standard, this concept is only one of many factors in a safeguard solution. I do not believe that these "standards" were conceived of as absolute design standards to be universally applied. They play appropriate roles in establishing an effective and cost-effective safeguards approach.

Repository

Until sealed, the repository is a facility under safeguards. Material once placed in the repository is no longer accessible. The regulations governing the repository require that it will not be sealed for 50 years after loading to provide assurance that nothing dangerous is happening. DOE has said that it will be conservative and wait 100 years.

These DOE containers will be surrounded by thousands of containers of spent fuel or HLW in the repository. Therefore radiation and geometry provide proliferation resistance. For assessment of proliferation risk, the canisters will be surrounded by material with the same or higher proliferation risk potential.

Self-Protection Criterion

The IAEA's Self-Protection Guideline is a definition, not a necessity for proliferation resistance. Meeting nonproliferation acceptability involves a set of safeguards technologies. Materials that are not regarded as "self-protecting" can successfully be safeguarded by many other means. Therefore, meeting a Self-Protection Guideline is not in itself a figure of merit for nonproliferation

Study requested is *An Assessment of Alternative Technologies without Processing--Needs Assessment of Processing Case for Comparison.* The EIS must include assessment of alternatives not restricted by the proposing organization. Alternatives currently in operation need to be assessed, even if different from national policy.

Current operations provide actual cost data for comparison. The discussion covered experience, problems and solutions. The base case data have less uncertainty than for the alternatives included in the assessment.

An Independent Study by DOE. The final environmental impact statement summary on policy concerning foreign research reactor spent fuel (DOE/EIS-0218F Feb. 1996, referred to as FEIS-96S) explains use of the EIS process, indicates that some spent fuel will be processed at SRL, and discusses (Summary p. 23) considerations that would affect decisions to process spent fuel.

The ROD on FRRFM and FEIS-96S call for an "independent study on the nonproliferation and other implications of reprocessing of spent fuel from foreign research reactors" to be initiated in mid-1996. The panel heard a presentation on DOE's plans for this effort and its progress to date. The study, however, is being done by DOE, and the presentation stated that the Office of Arms Control and Nonproliferation is conducting it. A mid-1998 target date was indicated.

Questions were raised about whether this approach could be justified as an independent study. Jon Wolfstal of DOE responded that one outside expert had prepared a preliminary draft and a second had reviewed and edited it. Despite statements of Mr. Wolfstal about following standard procedures for public comment and response, "process" does not produce credibility.

Publication of a draft for comment is used to assist the DOE in preparing the final report. The comments that are received are addressed in a section of the final report, and generally some brief discussion is included about how comments were handled and why. This particular approach to "process" does not afford opportunity for discussion or feedback between commentors and the authors, nor is there any independent arbiter. It is recognized that the authors are the ones responsible for the report, but it is seldom made clear when Department or White House policy sets the tone or the conclusions, and makes open discussion of certain comments impossible.

Since it is this same limited process that DOE intends to use in this "independent study," I am concerned that there will be no opportunity for open discussion or debate on the critical issue of evaluation of all alternatives including processing options and the implications of them. I have personally experienced frustration with the comment process on DOE decision documents. Therefore, I would be very dubious that the current approach would permit DOE to obtain the benefit of a truly independent study. A different approach may be advisable in resolving this issue.

Can an Adequate Record to Support this Contention be Provided for Assessment? As noted above, the ROD on FRRFM and the FEIS-96S call for an "independent study on the nonproliferation and other implications of reprocessing of spent fuel from foreign research reactors." This statement makes it obvious that a study should include processing options among the alternatives. However, this was specifically not done in the Task Team Report.

In my opinion, whether or not use of processing encourages plutonium processing in other nations is a matter of conjecture. I have observed that this opinion is widely shared both internationally and within the United States. Therefore, a discussion of this key point should be included for any technology option for which it is offered as a criterion for exclusion.

Countless precedents for this point can be found in commercial nuclear power plant licensing cases. EISs were rejected and returned to their sponsoring organizations for failure to adequately document their assessment of alternatives, even if there were special reasons why those alternatives might not be chosen. Many of the Environmental Impact Statements that were prepared for nuclear reactor construction permits were found to be "incomplete" by the Council on Environmental Quality in the early 1970s because they did not deal exhaustively with alternatives to the proposed nuclear power plants.

I made a presentation, along with Ruble Thomas of Southern Services and Charley Wylie of Duke Power Company, to Nuclear Regulatory Commission (USNRC) staff on the system planning concepts used by utilities in selecting among alternatives (about 1974; no reference available to me). We explained siting alternatives and fuel alternatives and the reasoning that was used to make successive decisions. The staff response was that these concepts were acceptable and logical, but that they should be described in environmental reports submitted by utilities and would be included in the environmental impact statements the USNRC would prepare. We pointed out that sometimes certain alternatives were impractical or unreasonable on their face, and argued that analyzing such alternatives should not be necessary in the EIS process, but we were advised to include all "reasonable" alternatives in our assessment.

Use of Processing for Waste Management

Processing has been used and will continue to be used as long as it is needed and facilities to do it are available for damaged or corroded spent fuel that might leak or cause contamination. However, DOE has taken the position that unless it is a health and safety matter, processing should not be used as a step in radioactive waste management.

A point was made that other nations may be considering use of reprocessing as a step in their waste management program. Obviously, this was once the case in the United States. In response to questions, Mr. Wolfstal mentioned France and Taiwan as nations that wish to do this. France does it commercially. Taiwan needs our consent to do it with United States origin fuel. These nations are nonproliferation treaty members and have stated that they will not use commercial reprocessing for weapons purposes.

It is certainly not clear to me that whether or not the United States uses processing to manage its assortment of DOE spent fuel and foreign research reactor spent fuel would be of any influence on these nations or any others.

Mr. Wolfstal pointed out that the French appeared to use the U.S. acceptance of MOX for disposition of excess weapons plutonium as an argument for why the rest of the world should continue to use civil plutonium. (A copy of the cited Newsletter from France on the French nuclear program is attached.)

Summary of Findings

Proliferation Resistance. There is no actual difference in proliferation resistance between the several alternatives considered in the Task Team report. The same would be true if the base case involving further use of processing in canyon facilities at Savannah River were chosen.

Following U.S. Policy. Where selection decisions are based on interpretation of U.S. policy rather than technical, cost-effectiveness and environmental merit, this must be explicitly stated and discussed in an EIS.

Follow Basic EIS Requirements. All realistic alternatives must be treated in an EIS.

International Implications. It is not credible to say that the choice of technology the United States makes for managing this spent fuel will have any effect on nonproliferation decisions other nations will make.

Independence. It is difficult to accept a report done by the nuclear nonproliferation of DOE as independent in such a controversial area.

Topic: Metallurgy and Corrosion
Consultant: Paul Shewmon

Introduction

The Westinghouse Savannah River (WSR) people are working only with Direct Disposal (D-D) and Melt-Dilute (M-D) processes for disposing of aluminum clad fuel so these are the only processes my comments will cover. The electrolytic process would use the M-D ingots as a uniform feed/electrode product. The WSR people agree that the aluminum clad fuel could also be reprocessed in the canyons at SR as long as these canyons are available. However, the return of fuel from research reactors could stretch on into the future until long after it would be economical to maintain the canyons in operation and then some other process like D-D or M-D would need to be used.

Processing

Direct Disposal. In this process the fuel element is dried; non-fuel-bearing material at the ends of the assemblies is removed; and the fuel-bearing material is sealed in a canister with dry air for shipment to interim storage and ultimately to a repository. The exact composition of the canister material has not yet been specified, but it will probably be an austenitic stainless steel. The compatibility of the fuel with the canister would not be a concern under repository conditions (temperatures) due to the inertness of the aluminum cladding and the stainless steel canister in dry air. It is expected that each fuel type can be put in a form acceptable for the repository. Requirements for the fuel form needed for a road-ready package and interim dry storage of aluminum-clad fuels received from basin storage have been recommended [1,2]. Also, the requirements for repository disposal of such fuel have been established and can be met [2].

Each fuel type will require some adjustment of the process to put in an acceptable form. The packing density of the fuel in the canister will be low due to the irregular shape of the fuel, and free space built into the subassembly to allow cooling water flow.

Melt-Dilute. Here the fuel assemblies will be dried and then melted in an induction furnace. The volatile fission product will be released at the melting temperature (850-1000 °C) and collected. Melting the fuel gives a large volume reduction and the number of canisters required will be reduced appreciably compared with the D-D product (by factors of about 4 x depending on the fuel). The cast product will have a low surface/volume, and the microstructure of the ingots will be more uniform than the heterogeneous D-D fuel. Two principal concerns with the D-D option are proliferation and criticality. M-D processing can remove both of these by dilution with unenriched uranium and aluminum. From a corrosion viewpoint, the M-D gives a product whose behavior is much more predictable in considerations of long term corrosion and whose composition could be optimized for such stability. One of the current research efforts of the WSR materials staff is study of the long term integrity of the waste form in water. In this vein they may have as a goal varying the composition to optimize the long term integrity of the waste on long term (10,000 yr.) exposure to water.

There are a variety of fuel types and geometries that must be handled in this program and these elements possess a limitless variety of histories. Characterizing these in enough detail to assure suitability for Direct Disposal can be time consuming, and melting is an excellent way to assure the uniformity of the product and reliability in processing.

The WSRC people have melted very few, if any, irradiated fuel elements, but they and others have melted a great deal of fuel for manufacturing Al-U fuel elements. Also, fission product release has been measured in severe accident studies [3]. Thus it would appear that the information needed for designing and building a facility for the melting and casting of these fuel elements is in hand and that the process could be put into operation with few if any surprises.

Dry Storage (Interim Storage). After the aluminum clad fuel has been processed for D-D or M-D it will be sealed in a canister with an inert atmosphere, which will probably be dried air. Aluminum forms a protective oxide film under these conditions and there would be virtually no measurable reaction of the waste form with the atmosphere or the canister for the years or decades that the waste may wait for placement in Yucca Mountain. The formation and stability of this oxide on various aluminum alloys is well established in the technical literature for temperatures near room temperature [4] and has been expanded to cover the product of the M-D process and higher temperatures by work at WSRC [5].

Answers to Assigned Questions

1. *Are DOE's plans for fuel handling, drying, etc., technically credible?* Yes. DOE's plans for fuel handling, drying, and interim storage are technically credible and the process steps are adequate to prevent significant fuel corrosion.

2. *For each of the processing options evaluated by DOE, are processing steps technically credible?* Yes, for the two processes under serious study, namely Direct Disposal and Melt Dilute. However, WSRC is not trying to prepare a basis for all of the processes considered in the report of the Research Reactor Task Team Study (Jack DeVine, Chairman).

3. *Would filling the canister void space with aluminum or some other sacrificial material make any difference in long-term corrosion of the waste form?* No, but it is quite likely that the cast product resulting from the Melt-Dilute process would have materially better corrosion resistance than that of the Direct-Disposal product. This is a topic currently under study.

4. *Will any of the waste forms resulting from any of the alternative processing operations be likely to increase internal corrosion of a standard repository container?* No.

5. *What is the status of R&D activities at Savannah River on the Melt-and-Dilute and Co-Disposal options? Are the R&D activities appropriately focused?* The metallurgical process information needed for the M-D and D-D processes is well in hand. The research activities needed for this have been well focused.

References

1. Sindelar, R.L., et al., "Acceptance Criteria for Interim Dry Storage of Aluminum-Alloy Clad Spent Nuclear Fuels," March 1966, WSRC-TR-95-0347.
2. "Alternative Aluminum Spent Nuclear Fuel Treatment Technology Development Status Report," October 1997, WSRC-TR-97-00345(U), Sec. 3.2-3.3.
3. Howell, J.P., "Fission Product Release from Spent Nuclear Fuel During Melting (U)", WSRC-TR-97-0112 (U).
4. Godard, H.P., "Oxide Film Growth Over 5 Years on Some Aluminum Sheet Alloys in Air of Varying Humidity at Room Temperature," J. Electrochem. Soc., 1967, v. 10, p. 354.
5. Lam, P.S., R.L. Sindelar, H.B. Peacock, Jr., Vapor Corrosion of Aluminum Cladding Alloys and Aluminum-Uranium Fuel Materials in Storage Environments, WSRC-TC-97-0120.

Topic: Cost and Schedule
Consultant: Richard I. Smith

In November of 1995, the Department of Energy (DOE) established the Research Reactor Spent Nuclear Fuel Task Team to assist in developing a technical strategy for interim management and ultimate disposition of the foreign and domestic aluminum-based research reactor spent nuclear fuel in DOE's jurisdiction, including both current inventory and expected receipts. The Task Team developed a two-volume report titled *Technical Strategy for the Treatment, Packaging, and Disposal of Aluminum-Based Spent Nuclear Fuel* [DeVine et al., 1996], issued in June of 1996. Subsequently, DOE contracted with the National Research Council (NRC) of the National Academy of Sciences to review the set of technologies evaluated in the Task Team report and suggest other alternatives that DOE might consider; to examine the waste package performance criteria developed by DOE for aluminum-based spent nuclear fuel and suggest other factors that DOE might consider; and to assess the cost and timing aspects of each of the disposition strategies proposed by DOE. To facilitate this review, the NRC assembled a team of experts in the fields of nuclear criticality, nuclear proliferation, cost and schedule, corrosion and metallurgy, processing and remote handling, and regulatory/waste acceptance. Copies of the Task Team report were provided to the experts selected to participate in the review, a two-day meeting was held in Augusta, Georgia on December 2 and 3, 1997, where the Task Team report was presented by its authors and additional presentations were made by various staff from the Savannah River Site (SRS) on progress toward implementation of the various strategies since the Task Team report was prepared.

Each of the groups of experts assembled by the NRC was posed a set of questions about the proposed strategies specific to its areas of expertise, to be answered from the information contained in the Task Team report, gathered at the Augusta meeting, and from any other sources available. This appendix is focused on the cost and schedule aspects of the problem.

General Comments

It was quite apparent from the strategies presented in the Task Team report that DOE's intent was to find ways to dispose of the aluminum-based spent fuel without recovering any of the residual highly enriched uranium from the spent fuel assemblies, presumably for reasons of non-proliferation. Reprocessing was not evaluated in the report and compared with the other alternatives, despite the fact that reprocessing of this type of fuel was presently ongoing and any comparison of alternatives should (must) include the possibility of continuing the current method of dealing with the spent fuel. As a result, the strategy with the highest probability of success, with the best-defined costs, and with a resultant waste product that is assured of repository acceptance, was not evaluated in the analyses.

In a subsequent report [Krupa 1997], several strategies have been devised that include reprocessing of the current inventory of spent fuel through about 2010 and applying some other treatment process to those fuels that enter the inventory in later years. As might be expected, those strategies result in completing the disposition of the bulk of the anticipated inventory of aluminum-based fuels in the least time, with the least cost, and the highest probability of success.

Specific Comments

Six questions about the Task Team report were posed to the cost and schedule experts by the Principal Investigator for this study. Each question is presented and this reviewer's responses and discussions are given in subsequent subsections.

Question 1: Are the cost data provided by DOE reasonably complete and transparent?

Response: In general, the answer is yes. The detailed costs associated with each strategy are presented in Appendix C of Volume II of the Task Team report. The costs are broken down sufficiently far to see which elements are important to the result and which elements are common to all strategies. Generally, the bases (sources) for the various cost elements are given, and the rationales for various assumptions made are also given. While one might disagree with some of the assumptions or cost values, those used are well documented. The data presented represent the state of knowledge at the time of the report. However, some of that data has been superseded by more recent cost evaluations [Krupa 1997].

Question 2: Are the cost and schedule estimates developed by DOE for the alternative processing options suitable as a basis for comparison and selection of one or more preferred alternatives?

Response: Yes and No. For those alternatives considered in the report, the data presented are probably sufficient for comparisons to be made and to select one or more preferred alternatives. The cost bases are generally internally consistent across the alternatives, the processes of each alternative are examined in sufficient detail to assure that no major cost elements have been overlooked. However, because continued reprocessing was not included in the analyses, there is no real basis for comparison between current practice and future possibilities.

There is an old axiom in the cost estimation business: "The less you know about a given process, the cheaper and easier it appears." Some of that phenomenon has likely occurred in the estimates for those processes for which little or no development or demonstration work has been carried out. Some of the uncertainty estimates for certain aspects of some alternatives seem rather large, but they may only reflect the state of knowledge at the time of the report. For example, the uncertainties assigned to the electrometallurgical (EM) treatment process are much larger than all processes except the GMODS process. The basic EM process had been demonstrated for other types of spent uranium fuel. Since that time, lab-scale development testing for the more complicated aluminum-removal process has been completed, and the developers are ready to proceed to engineering-scale development testing [Slater 1997]. Thus, confidence in success in developing the Al-U process would appear to have increased and the uncertainty in project costs would appear to have decreased, relative to those processes in the Task Team report that have not been demonstrated.

It is not obvious that the schedules contained in the Task Team report are achievable. In general, the key milestones were established by

DOE, without a bottom-up examination of the program elements necessary to reach those milestones. No consideration was given to the time required to establish a line-item in the DOE budget for construction of any new facilities nor to the time required to select contractors for design and construction of those facilities. As a result, most of the schedules are optimistic by several years as a minimum. Since delays in construction and operation of the required new facilities will require extended utilization of currently used water basins, total program costs will increase for each year of delay. Similarly, some of the processes have had little or no development work done. Any delays in developing and implementing the processes will also delay the program, with attendant cost increases. These types of delays may affect some alternatives more than others.

Question 3: Are the cost and schedule estimates developed by DOE for the alternative processing options suitable for budget planning purposes?

Response: No. The milestones artificially imposed on the Task Team considerations preclude using those schedules for developing budget estimates. They ignore the time necessary to place a project into the DOE line-item budget and the time (and money) necessary to select an architect-engineer and a construction contractor. They also ignore the time (and money) required to prepare and issue an environmental impact statement, or an environmental assessment, if such are necessary for these projects, and ignore the time (and money) needed to deal with satisfying Nuclear Regulatory Commission reviews and possible licensing of any new facilities or processes. At least several years would be added to the schedules outlined in the Task Team report, and extending the period during which the wet basins are needed for storage and handling of spent fuel will also add significantly to the overall project life-cycle cost. While these schedule extensions will increase the cost of the proposed projects, they are generally common to all alternatives (except perhaps continued reprocessing) and would not significantly affect the comparison between the alternatives presented in the Task Team report.

The most recent cost analyses for the proposed alternatives [Krupa 1997] do take into account at least most of the above-described schedule delays and are more nearly suitable for preparing long-range budget estimates.

Question 4: Has DOE considered the costs of program delays in its budget development or budget planning for this program?

Response: No and Yes. The effects of program delays were not seriously considered in the Task Team report [DeVine et al., 1996], since such delays were generally common to all alternatives and did not affect the comparisons. The extended schedules considered in the most recent program cost analyses [Krupa 1997] are reflected in the projected program costs. However, no costs are included to reflect further technical development efforts on undemonstrated technologies. Apparently, these types of activities are being funded from other sources. Also, no schedule allocations are made to accommodate such development efforts. Any technical difficulties in proving out a selected treatment process could result in additional schedule delays.

Question 5: Are the cost and schedule estimates for implementing the alternative processing options consistent with DOE procedures and systems? If not, has DOE identified what changes must be made to achieve its cost and schedule targets?

Response: The first part of this question is essentially a restatement of Question 3 and is discussed there. The response to the second part of the question is not clear. Apparently DOE has not yet decided how to fund the project, either by privatization or by the budget line-item project approach. Both approaches require a significant amount of lead time to establish the appropriate contractual arrangements with contractors. To achieve the rather short schedules currently proposed, the project will have to be highly organized and tightly controlled, with the authority to make necessary decisions held at the local (site) level. It is not clear that DOE has yet made the decisions necessary to allow the project to go forward in an optimum fashion nor that it will make those decisions any time soon.

Question 6: Are the cost and schedule milestones that are laid out in the Research Reactor Task Force Report for selecting and implementing an alternative processing option being met?

Response: Difficult to Predict. The schedule for alternative selection in the Task Team report called for a decision late in 1999. Some winnowing of the alternatives originally recommended for further study has already occurred in that development activities on the press and dilute option have essentially been suspended for lack of funding. Development on the melt and dilute option is progressing reasonably well. Lab-scale development for the EM process has been completed and engineering-scale development is scheduled to start soon. It is not yet clear that the various co-disposal approaches will be able to qualify for repository acceptance, so those alternatives may be in doubt. The initial 6-8 years or so of the three reprocessing alternatives suggested by Krupa [1997] obviously can be implemented as quickly as space is available in the H-Canyon reprocessing schedule, although exactly which process should be utilized for the low-throughput period following closure of the H-Canyon reprocessing facility in 2010 is not clear. One possibility not yet considered by DOE would be to install a relatively low-throughput aluminum-removal stage of the EM process in the same hot cell that is presently occupied by the EM process being used currently for EBR-II fuel at INEEL. The uranium feed stream from the aluminum-removal step would feed directly into the existing uranium refining process to complete the separation of the uranium from the residual fission products. This approach would avoid the construction of any new facilities at all and require only addition of the incremental equipment for the aluminum-removal step to the existing hot cell system. However, for best economics, an ongoing mission for the existing uranium EM process would be needed (e.g., treatment of the N-Reactor fuel from Hanford prior to repository disposal), because the cost per unit of fuel processed for facility operations might be rather high if only the aluminum-based fuel stream were being processed, because the facility would have to be maintained ready for service even when there was no aluminum-based fuel in inventory.

Other Comments

The most recent analyses [Krupa 1997] show that the continued reprocessing of the aluminum-based spent fuel in the H-Canyon facility at Savannah River Site is the most cost-effective approach and can eliminate the existing inventory from both SRS and INEEL in the least time, with the greatest certainty of success (i.e., a guaranteed repository-acceptable waste form) and recovery of a valuable resource (highly enriched ^{235}U) to be blended down for future use in our domestic nuclear power industry. Unfortunately, DOE has shown a tendency in the past to prematurely close existing reprocessing facilities before their missions were complete, apparently for the purpose of satisfying some non-proliferation policy desires and to gain the approval of those members of the public who are opposed to nuclear power in general and to reprocessing in particular. As a result of such premature shutdowns at INEEL and at Hanford, DOE now has a large inventory of residual aluminum-based spent fuel stored at INEEL and a large inventory (about 2,300 tons) of spent metallic uranium fuel from the final years of N-Reactor operation stored at Hanford in wet pools where it is slowly corroding into sludge. Because of the safety implications of a pool leaking into the Columbia River, DOE has had to establish a major program, which has been underway for the past five years or so to remove this fuel from the pools and place it into dry storage away from the river. The most recent project cost estimate is now $1.08 billion, with completion still several years away. The final product of this project will be metallic fuel elements stored in steel canisters, a product unlikely to be acceptable to the repository without further treatment before disposal, so the total cost of preparing this material for disposal will certainly exceed the current estimate by a significant amount.

Continuing to operate the Hanford reprocessing facility (PUREX) instead of shutting it down, and reprocessing all of that material into separated fuel material and fission product wastes would have cost about $300 million to $400 million and required about 3 years of operation. Contrasting those fairly well-known costs and schedule with the presently estimated (and still uncertain) cost of $1.08 billion over 7-8 years for the current project suggests that the decision to close PUREX before its mission was completed was a major mistake.

DOE is again faced with making decisions related to the aluminum-based fuel disposition program that are similar to the PUREX and INEEL decisions (i.e., to shut down an existing reprocessing capability before the mission has been completed to satisfy some policy desires related to non-proliferation or to continue reprocessing until the inventory of aluminum-based spent fuel has been eliminated). All of the analyses to date show that continued reprocessing is the best, fastest, cheapest, and most certain of success of all of the alternatives considered. I trust DOE will not allow the somewhat tenuous non-proliferation policy considerations to reject the path forward that is technically and economically the best.

References

Bailey, R.W., and M.S. Gerber, Purex/UO₃ Facilities Deactivation Lessons Learned History, HNP-SP-1147, Rev 2, Fluor-Daniel Hanford Company, Richland, Washington, 1997.

DeVine, J. C., et al., Technical Strategy for the Treatment, Packaging, and Disposal of Aluminum-Based Spent Nuclear Fuel, a report of the Research Reactor Spent Nuclear Fuel Task Team. U. S. Department of Energy, Washington, D.C., June 1996.

Krupa, J. L., Savannah River Site Aluminum-Clad Spent Nuclear Fuel Alternative Cost Study, *Rev 1(U)*, WSRP-RP-97-299 REV. 1, Westinghouse Savannah River Company, Aiken, South Carolina, December 1997.

Slater, S. A., and J. L. Willit. Electrometallurgical Treatment of Aluminum-Based Fuel, presented at the Augusta, Georgia review meeting, December 2-3, 1997.

Westinghouse Hanford Company, PUREX/UO₃ Standby Management Plan, WHC-SP-0631, Rev. 1, 1991.

APPENDIX E

Biographical Sketches of Consultants

Harold M. Agnew served as president of General Atomics and director of Los Alamos National Laboratory before his retirement and is now an adjunct professor at the University of California, San Diego. His primary interests relate to nuclear physics and its application to defense, energy, and biological sciences. He worked with Enrico Fermi on the first nuclear chain reaction and assisted in the development of the atomic bomb. He has served on the White House Science Council and currently serves on General Atomic Company's advisory board. Dr. Agnew has received numerous awards, including the Fermi Award from DOE for his contributions to nuclear physics, and he is a member of the National Academy of Sciences and National Academy of Engineering. He received an A.B. degree from the University of Denver and M.A. and Ph.D. degrees from the University of Chicago.

John F. Ahearne is the director of the Sigma Xi Center for Sigma Xi, The Scientific Research Society, a lecturer in public policy and adjunct professor in civil and environmental engineering at Duke University, and an adjunct scholar at Resources for the Future. His professional interests are reactor safety, energy issues, resource allocation, and public policy management. He has served as commissioner and chairman of the U.S. Nuclear Regulatory Commission, systems analyst for the White House Energy Office, Deputy Assistant Secretary for Energy, and Principal Deputy Assistant Secretary for Defense. Dr. Ahearne currently serves on the Department of Energy's Environmental Management Advisory Board and the National Research Council's Board on Radioactive Waste Management. He is a fellow of the American Physical Society, American Association for the Advancement of Science, and American Academy of Arts and Sciences, and he is a member of the National Academy of Engineering. He received his B.S. and M.S. degrees from Cornell University and his Ph.D. degree in physics from Princeton University.

211

Francis M. Alcorn is manager of nuclear criticality safety with BWX Technologies in Lynchburg, VA. His professional interests include criticality evaluations and calculations, nuclear safety audits, nuclear fuel costs, and critical experiments analysis. He is a member of the Executive Committee on Nuclear Criticality Safety and the ANS-8 of the American Nuclear Society, which writes American National Standards for nuclear criticality safety. He also has served as past chairman of the society's Nuclear Criticality Safety Division. He received a B.S. in nuclear engineering from North Carolina State University and an MBA from Lynchburg College.

Maurice W. Angvall served as Bechtel's manager of estimating for the U.S. Department of Energy Savannah River Operations Office and for Bechtel National Inc. in San Francisco prior to his retirement. He has 38 years of experience in construction management, design engineering, project control, estimating, and financial analysis on various industrial and nuclear facilities and defense related engineering and construction projects. He received his B.S. degree in civil engineering from the University of Minnesota and his MBA in finance from the University of California at Berkeley.

Robert M. Bernero recently retired from 23 years of service with the U.S. Nuclear Regulatory Commission (USNRC), where he held numerous positions up to director of the Office of Nuclear Material Safety and Safeguards. Prior to joining the USNRC, he worked for the General Electric Company in nuclear technology. He currently consults on nuclear safety related matters, and serves as a member of the Commission of Inquiry for an international review of Swedish nuclear regulatory activities. His professional expertise includes licensing, inspection, and environmental review of industrial, medical, academic, and commercial uses of radioisotopes. He received a B.A. degree from St. Mary of the Lake (Illinois), a B.S. degree from the University of Illinois, and an M.S. degree from Rensselaer Polytechnic Institute.

Joseph S. Byrd is distinguished professor emeritus of electrical and computer engineering and the University of South Carolina and an expert consultant in robotics for the International Union of Operating Engineers National Hazmat Program. Prior to joining the faculty at South Carolina, he managed the robotics technology group and the engineering development Group at the DuPont Savannah River Laboratory. His professional interests include robotics development and the application of robotics to environmental remediation. He has received several professional awards including the J.W. Lathrop Oustanding Electrical Engineering Educator Award, the Ray Goertz Award, and the Samuel Litman Distinguished Engineering Professor of Engineering award. He received a B.S. degree from Clemson University and a M.S. degree from the University of South Carolina, both in electrical engineering.

Robert L. Dillon served in various management assignments at the U.S. Department of Energy's Richland Operations Office at Hanford until his retirement. His professional interests include corrosion product transport in aqueous, organic, and liquid metal coolants, stress corrosion, and decontamination. He received a B.A. degree in chemistry from Reed College, and M.S. and Ph.D. degrees, both in physical chemistry, from Northwestern University.

G. Brian Estes is the former director of construction projects, Westinghouse Hanford Company, where he managed construction projects in support of operations and environmental cleanup of the Department of Energy Hanford nuclear complex. Prior to joining Westinghouse, Mr. Estes completed 30 years in the Navy Civil Engineer Corps, achieving the rank of rear admiral. He is a registered professional engineer in Illinois, served on the executive committee of the Construction Industry Institute, and is former national director of the Society of American Military Engineers. He holds a B.S. degree in Civil Engineering from the University of Maine and an M.S. degree in Civil Engineering from the University of Illinois.

Harry D. Harmon has over 25 years experience with Dupont and Westinghouse at Department of Energy's Savannah River and Hanford sites in waste management and radiochemical processing activities. For the past seven years, he has focused almost entirely on high level waste operations and related technology development. His expertise includes nuclear fuel reprocessing and recovery and purification of uranium, plutonium, and transplutonium products. He also has significant experience at Savannah River Laboratory managing process technology development as well as analytical development and environmental technology. He holds a B.S. in chemistry from Carson-Newman College and a Ph.D. degree in inorganic and nuclear chemistry from the University of Tennessee, Knoxville.

Valerie L. Putman is a senior engineer of criticality safety for Lockheed Martin Idaho Technologies Company. Her experience and interests lie in nuclear criticality safety and in accident and incident investigation. Ms. Putman is an active member of the American Nuclear Society and its Nuclear Criticality Safety, Human Factors and Environment Divisions and Professional Development Coordinating Committee. She is also active in several American Nuclear Society ANS-8 standards writing groups on criticality safety. Ms. Putman received a B.S. in both applied physics and mathematics from the University of Utah and an M.E. in mechanical engineering with nuclear emphasis from the University of Idaho. She is pursuing her Ph.D. in nuclear engineering through a joint program with Idaho State University and the University of Idaho.

A. David Rossin is Center Affiliated Scholar at the Center for International Security and Arms Control at Stanford University, and is writing a book about the origin of U.S. nonproliferation policies. He is President of Rossin and Associates and consultant to Lawrence Livermore National Laboratory and Los Alamos National Laboratory on nuclear and energy technologies, nonproliferation and waste management. Dr. Rossin previously served as Assistant Secretary for Nuclear Energy, U.S. Department of Energy, and as president of the American Nuclear Society. He received a B.S. degree in engineering physics from Cornell University; an M.S. degree in nuclear engineering from the Massachusetts

Institute of Technology; an MBA from Northwestern University; and a Ph.D. in metallurgy from Case Western Reserve University.

Paul G. Shewmon retired recently as Professor at the Ohio State University and as a member of the Nuclear Regulatory Commission's Advisory Committee on Reactor Safeguards. His expertise is in materials sciences, especially the structure and composition of metals and alloys. Dr. Shewmon previously served as director of the Materials Science Division of the U.S. National Science Foundation and the director of the Metallurgy and Materials Division of Argonne National Lab. He received a B.S. degree in metallurgical engineering from the University of Illinois, and M.S. and Ph.D. degrees, also in metallurgical engineering, from the Carnegie Institute of Technology. Dr. Shewmon is a member of the National Academy of Engineering.

Richard I. Smith retired from Battelle's Pacific Northwest Laboratories in 1996 after nearly 40 years of scientific activities on the Hanford Site. During that time he conducted experiments on and performed analyses of reactor neutronics for plutonium-uranium fueled reactors, and from 1978 until his retirement led a multi-year program of studies that estimated the costs, radiation doses, and waste volumes associated with the decontamination and decommissioning of licensed commercial nuclear facilities, in support of the U.S. Nuclear Regulatory Commission. He also led studies to select concepts for DOE's MRS program and participated in the initial conceptual design of the MRS facility. He has participated in the development of a number of technical documents for and has acted as a consultant to the International Atomic Energy Agency in the areas of spent fuel management and reactor decommissioning. He received a B.S. in Physics from Washington State University and a M.S. in Applied Physics from University of California at Los Angeles, and is a licensed professional engineer in nuclear engineering in the states of California and Washington.

APPENDIX F

Documents Received in This Study

Argonne National Laboratory. 1963. Description and Proposed Operations of Fuel Cycle Facility for the Second Breeder Reactor (EBR-II). ANL-6605. Chicago, Illinois: Argonne National Laboratory.

Civilian Radioactive Waste Management System. 1997. Total System Performance Assessment Sensitivity Studies of U.S. Department of Energy Spent Nuclear Fuel. A00000000-01717-5705-00017, Rev. 01. September 30, 1997. Las Vegas, Nevada: TRW Environmental Safety Systems, Inc.

Civilian Radioactive Waste Management System. 1997. Evaluation of Codisposal Viability for Aluminum-Clad DOE-Owned Spent Fuel: Phase 1. Intact Codisposal Canister. BBA000000-01717-5705-00011, Rev. 01. August 15, 1997. Las Vegas, Nevada: Civilian Radioactive Waste Management System.

Forsberg, Charles. Oak Ridge National Laboratory. 1997. Letter to Kevin Crowley with attachments on the use of excess depleted uranium to improve disposal of spent nuclear fuel, dated December 16, 1997. Oak Ridge, Tennessee: Oak Ridge National Laboratory.

Howell, J. P. 1997. Fission Product Release From Spent Nuclear Fuel During Melting (U). WSRC-TR-97-0112 (U). Aiken, South Carolina: Westinghouse Savannah River Company.

INEEL Spent Nuclear Fuel Task Team. 1997. Technical Strategy for the Management of INEEL Spent Nuclear Fuel. March 1997. Idaho Falls, Idaho: U.S. Department of Energy, Idaho National Engineering and Environmental Laboratory.

Jonke, A. A. 1965. Process studies on the recovery of uranium from highly enriched uranium alloy fuels. Argonne National Laboratory

Chemical Engineering Division Semiannual Report, January-June 1965. ANL-7055. Chicago, Illinois: Argonne National Laboratory.

Lam, P. S., R. L. Sindelar, and H. B. Peacock, Jr. 1997. Vapor Corrosion of Aluminum Cladding Alloys and Aluminum-Uranium Fuel Materials in Storage Environments (U). WSRC-TR-97-0120. Aiken, South Carolina: Westinghouse Savannah River Company.

Large, W. S. and R. L. Sindelar. 1997. Review of Drying Methods for Spent Nuclear Fuel (U). WSRC-TR-97-0075 (U). Aiken, South Carolina: Westinghouse Savannah River Company.

Louthan, Jr., M. R. 1996. Task Plan for Engineering Test Protocol for Metallic Waste Forms (U). SRT-MTS-96-2064. December 31, 1996. Aiken, South Carolina: Savannah River Technology Center.

McKibben, J. M., Westinghouse Savannah River Company. 1997. Letter to Milton Levenson on aluminum spent fuel disposition, dated January 5, 1997. Aiken, South Carolina: Westinghouse Savannah River Company.

National Research Council. 1995. An Assessment of Continued R&D into an Electrometallurgical Approach for Treating DOE Spent Nuclear Fuel. Washington, D.C.: National Academy Press.

Nuclear Waste Technical Review Board. 1995. Report to the U.S. Congress and the Secretary of Energy. Arlington, Virginia: Nuclear Waste Technical Review Board.

Nuclear Waste Technical Review Board. 1997. Handouts and transcript on diskette from December 17, 1997 meeting in Atlanta, Georgia.

Office of the Press Secretary, The White House. 1993. Press Release/Fact Sheet: Nonproliferation and Export Control Policy. Washington, D.C.: The White House.

Peacock, H. B. 1997. Task Plan for Development of Dilution Technologies for Aluminum-Base Spent Nuclear Fuel (U). SRT-MTS-96-2063. January 27, 1997. Aiken, South Carolina: Savannah River Technology Center.

Research Reactor Spent Nuclear Fuel Task Team. 1996. Technical Strategy for the Treatment, Packaging, and Disposal of Aluminum-

Based Spent Nuclear Fuel. Prepared for U.S. Department of Energy Office of Spent Fuel Management (2 volumes).

Sindelar, R. L. 1996. Plan for Development of Technologies for Direct Disposal of Aluminum Spent Nuclear Fuel (U). SRT-MTS-96-2047, Rev. 1. Aiken, South Carolina: Savannah River Technology Center.

Sindelar, R. L., H. B. Peacock, Jr., P. S. Lam, N. C. Iyer, and M. R. Louthan, Jr. 1996. Acceptance Criteria for Interim Dry Storage of Aluminum-Alloy Clad Spent Nuclear Fuels (U). WSRC-TR-95-0347 (U). Aiken, South Carolina: Westinghouse Savannah River Company.

Skidmore, Eric T. 1997. Task Plan for Characterization of DOE Aluminum Spent Nuclear Fuel (U). SRS-MST-97-2004. January 31, 1997. Aiken, South Carolina: Savannah River Technology Center.

Slater, Susan and James Willit. 1997. Electrometallurgical Treatment of Aluminum-Based Fuel. Viewgraphs provided at the December 2-3, 1997 meeting. Chicago, Illinois: Argonne National Laboratory.

TRW Environmental Safety Systems, Inc. 1995. Preliminary Requirements for the Disposition of DOE Spent Nuclear Fuel in a Deep Geologic Repository. DI: A00000000-00811-1708-00006 REV 00. December 15, 1995. Vienna, Virginia: TRW Environmental Safety Systems, Inc.

TRW Environmental Safety Systems, Inc. 1997. Mined Geologic Disposal System Waste Acceptance Criteria. B00000000-01717-4600-00095 REV 00. September 1997. Las Vegas, Nevada: TRW Environmental Safety Systems, Inc.

TRW Environmental Safety Systems, Inc. 1997. OCRWM Data Needs for DOE Spent Nuclear Fuel. DI: A00000000-01717-2200-00090 Rev 02. September 1997. Vienna, Virginia: TRW Environmental Safety Systems, Inc.

U.S. Department of Energy. 1992. Task Force Study on DOE Spent Fuel Reprocessing, Summary Report (Predecisional Draft). February 1992. Washington, D.C.: U.S. Department of Energy.

U.S. Department of Energy. 1992. Memorandum from DOE Secretary James D. Watkins entitled, "Highly Enriched Uranium Task Force

Report". February 24, 1992. Washington, D.C.: U.S. Department of Energy.

U.S. Department of Energy. 1992. Claytor, Richard A., Assistant Secretary for Defense Programs. Memorandum to the Secretary of Energy entitled "A Decision on Phaseout of Reprocessing at the Savannah River Site (SRS) and the Idaho National Engineering Laboratory (INEL) is Required." April 28, 1992. Washington, D.C.: U. S. Department of Energy.

U.S. Department of Energy. 1995. Programmatic Spent Fuel Management and Idaho National Engineering Laboratory Environmental Restoration and Waste Management Programs Final Environmental Impact Statement. Idaho National Engineering Laboratory, DOE/EIS-0203-F (14 volumes). Idaho Falls, Idaho: U.S. Department of Energy.

U.S. Department of Energy. 1996. Memorandum from DOE Deputy Assistant Secretary Jill E. Lytle entitled, "Approval of Path Forward for the Management of Aluminum-Based Research Reactor Spent Nuclear Fuel at the Savannah River Site". Washington, D.C.: U.S. Department of Energy.

U.S. Department of Energy. 1996. Integrated Data Base Report—1995: U.S. Spent Nuclear Fuel and Radioactive Waste Inventories, Projections, and Characteristics. DOE/RW-0006, Rev. 12. Washington, D.C.: Office of Environmental Management.

U.S. Department of Energy. 1996. Final Environmental Impact Statement: Proposed Nuclear Weapons Nonproliferation Policy Concerning Foreign Research Reactor Spent Nuclear Fuel, Summary. DOE/EIS-0218F. February 1996. Washington, D.C.: U.S. Department of Energy.

U.S. Department of Energy. 1996. DOE-Owned Spent Nuclear Fuel Technology Integration Plan. DOE/SNF/PP-002 REV 1. May 1996. Idaho Falls, Idaho: U.S. Department of Energy.

U.S. Nuclear Regulatory Commission. 1997. Code of Federal Regulations, Title 10, Part 60, Disposal of High-Level Radioactive Wastes in Geologic Repositories. Washington, D.C.: U.S. Government Printing Office.

Westinghouse Savannah River Company. 1996. Savannah River Site FY97 Spent Nuclear Fuel Interim Management Plan (U). WSRC-RP-96-530. October 1996. Aiken, South Carolina: Westinghouse Savannah River Company.

Westinghouse Savannah River Company. 1997. Alternative Aluminum Spent Nuclear Fuel Treatment Technology Development Status Report (U). WSRC-TR-97-0084. April 1997. Aiken, South Carolina: Westinghouse Savannah River Company.

Westinghouse Savannah River Company. 1997. Alternative Aluminum Spent Nuclear Fuel Treatment Technology Development Status Report (U). WSRC-TR-97.00345(U). October 1997. Aiken, South Carolina: Westinghouse Savannah River Company.

Westinghouse Savannah River Company. 1997. Savannah River Site Aluminum-Clad Spent Nuclear Fuel Alternative Cost Study (U). WSRC-RP-97-299 Rev. 1. December 1997. Aiken, South Carolina: Westinghouse Savannah River Company.

Westinghouse Savannah River Company. 1998. Savannah River Site FY 1998 Spent Nuclear Fuel Interim Management Plan. WSRC-RP-97-00922. January 1998. Aiken, South Carolina: Westinghouse Savannah River Company.

Handouts from the November 4-5, 1997 Meeting in Aiken, South Carolina:

1. Overview of Aluminum-Based Spent Nuclear Fuel Management at the Savannah River Site (Karl Waltzer, DOE-Savannah River).

2. Overview of Alternative Spent Fuel Disposition Technologies for National Academy of Sciences (Karl Waltzer, DOE-Savannah River).

3. Waste Acceptance Criteria, Aluminum SNF Forms (Natraj Iyer, WSRC).

4. Development of Direct/Co-Disposal Technology for Aluminum-Based Spent Nuclear Fuel (R.L. Sindelar, WSRC).

5. Overview of Alternate Technology Program for National Academy of Sciences (Mark Barlow, WSRC).

6. Alternative Technology Cost Development (J. F. Krupa, WSRC).

7. Melt-Dilute Process Technology for Spent Nuclear Fuel Storage (H. B. Peacock, WSRC).

8. SRS Material Stabilization Strategy, October 1997 Schedule (John Dickenson, WSRC).

Handouts from the December 2-3, 1997 Meeting in Augusta, Georgia:

1. Overview of Aluminum-based Spent Nuclear Fuel Management at the Savannah River Site (Karl Waltzer, DOE-Savannah River).

2. Technical Strategy for the Treatment, Packaging, and Disposal of Aluminum-Based Spent Nuclear Fuel Volume 1. June 1996.

3. Research Reactor SNF Task Team (Jack DeVine, Polestar).

4. Alternative Technology Program—Progress and Path Forward (Mark Barlow, WSRC).

5. Acceptability of Waste Forms: NAS Review of Al-Based SNF Alternative Technology Selection (Hugh A. Benton, TRW).

6. Waste Form and Co-Disposal Waste Package for Aluminum-Based Research Reactor Fuel: NAS Review of Al-Based SNF Alternative Technology Selection (Hugh A. Benton, TRW).

7. Alternate Technologies Process Descriptions (J. R. Murphy, WSRC).

8. Metallurgy and Corrosion Issues Aluminum SNF Disposition (Natraj Iyer, WSRC).

9. Disposal Criticality Analysis for Al-based DOE Fuel—MIT and ORR SNF (Peter Gottlieb, TRW).

10. Alternative Technology Cost Development (J. F. Krupa, WSRC).

11. Nonproliferation Study of Research Reactor Spent Fuel Management Alternatives (Jon Wolfstal, DOE Office of Nonproliferation).

12. Alternative Aluminum SNF Treatment Technology Path Forward (Natraj Iyer, WSRC).

APPENDIX G

Acronyms and Definitions

Aluminum Spent Nuclear Fuel. Irradiated fuel that contains uranium-aluminum matrix fuel elements and (or) is clad in aluminum.

Austenitic Stainless Steel. Nickel-chromium stainless steel identified as 300 series.

Canyon. Facility used to reprocess spent nuclear fuel, so named because on its long, narrow shape. There are two operating reprocessing facilities at Savannah River, the F Canyon and H Canyon.

CFR. Code of Federal Regulations.

Chloride Volatility Treatment. A process for treating aluminum spent fuel that involves the reaction of the spent fuel with chlorine or HCl gas at elevated temperatures to produce volatile chlorides, which are subsequently recovered by scrubbing and fractional distillation.

Cladding. A thin metal covering on a fuel element comprised of alloys such as aluminum, zircaloy, or stainless steel.

Conventional Reprocessing. A solvent extraction process for separating and recovering uranium and, if desired, plutonium from spent nuclear fuel.

Cost Comparison Point. The estimated cost for each of the treatment options evaluated by the Task Team.

Cost of Time. Operational costs of a facility that are unrelated to actual production activities, including management and administrative costs, costs of supporting workers in a stand-by mode, and other operational costs that are time related rather than production or throughput related.

Criticality Event. A self-sustaining nuclear reaction like that which occurs in a nuclear reactor.

CRWMS. Civilian Radioactive Waste Management System.

223

D&D. Decontamination and Decommissioning.

Depleted Uranium. Uranium that is depleted in uranium-235 relative to natural abundances.

Direct Co-Disposal Treatment. A process for treating aluminum spent fuel that involves drying the fuel and placing it into a canister for shipment to the repository for loading into a repository container with other canisters of vitrified waste.

Disposable Canister. A stainless steel canister whose primary purpose is to protect the spent fuel or the treated equivalent during interim storage, shipping, and handling operations.

Disposal Container. A container consisting of corrosion-resistant metallic layers designed to hold a number of spent fuel assemblies or high-level waste glass logs for disposal in a repository. This container is expected to maintain its integrity for thousands of years.

Dissolve and Vitrify Treatment. A process for treating aluminum spent fuel that involves dissolution of the fuel in acid along with depleted uranium to reduce the uranium-235 concentration to 20 percent or less by mass and vitrification of the resulting liquid waste stream.

DOE. U.S. Department of Energy.

DOE-Savannah River. U.S. Department of Energy, Savannah River Field Office and Westinghouse Savannah River Company staff.

DOE-Yucca Mountain. U.S. Department of Energy, Office of Civilian Radioactive Waste Management and its management and operating contractor.

DRR. Domestic Research Reactor.

DWPF. Defense Waste Processing Facility, a facility located at Savannah River for vitrifying high-level waste from defense operations.

EIS. Environmental Impact Statement.

Electrometallurgical Treatment. A process for treating aluminum spent fuel that involves melting and electrorefining to separate aluminum, uranium, and fission products.

EM. U.S. Department of Energy, Office of Environmental Management.

EPA. U.S. Environmental Protection Agency.

Fissile Isotope. An isotope that will fission in the presence of low-energy (thermal) neutrons, for example, uranium-235 and plutonium-239 (see, for example, *A Guide to Nuclear Power Technology,* John Wiley and Sons).

Fission. A process involving the separation of the nucleus of an atom into two (and sometimes three) fragments, accompanied by the release of neutrons and energy.

Frit. Powdered borosilicate glass used in the vitrification process.

FRR. Foreign Research Reactor.

Gaseous Fission Products. Isotopes produced by fission of uranium and plutonium that exist in a gaseous state at room temperature and pressure, for example, krypton and zenon.

Glass Material Oxidation and Dissolution Treatment. A process for treating aluminum spent fuel that involves melting the spent fuel with depleted uranium, adding lead oxide to oxidize the metals, and then adding frit to make glass with a uranium-235 enrichment of 20 percent or less by mass.

GMOD. Glass Material Oxidation and Dissolution Treatment.

HEU. Highly Enriched Uranium, uranium that contains more than 20 percent uranium-235 by mass.

HFIR. High Flux Isotope Reactor, a research reactor located at the Oak Ridge site in Tennessee.

HLW. High-Level Waste, the liquid by-product of conventional reprocessing, which contains fission products and trace amounts of uranium and plutonium.

Hot Cell. A physically isolated and heavily shielded space in which highly radioactive materials can be handled by remote control.

IAEA. International Atomic Energy Agency.

INEEL. Idaho National Engineering and Environmental Laboratory, formerly the Idaho National Engineering Laboratory (INEL).

LEU. Low-Enriched Uranium, uranium that contains 20 percent or less uranium-235 by mass.

Life-Cycle Costs. The total costs of designing, constructing, operating, and decontaminating and decommissioning a treatment or storage facility.

Melt and Dilute Treatment. A process for treating aluminum spent fuel that involves melting the spent fuel along with depleted uranium to produce an alloy that has a uranium-235 enrichment of 20 percent or less by mass.

MGDS. Mined Geological Disposal System.

MTHM. Metric Tons Heavy Metal, the amount of heavy metal (uranium, thorium, and plutonium) present in fresh (unirradiated) fuel.

Neutron Poison. A neutron absorbing material such as boron that can be incorporated into the spent fuel storage and shipping canister to reduce the likelihood of a criticality event.

NRC. National Research Council.

OCRWM. U.S. Department of Energy, Office of Civilian Radioactive Waste Management.

Off-Gas Treatment. A process for capturing gaseous or volatile fission products that are released when spent fuel is melted.

OMB. Office of Management and Budget.

PA. Performance Assessment.

PEIS. Programmatic Environmental Impact Statement.

P.I. Principal Investigator.

P.L. Public Law.

Plasma Arc Treatment. A process for treating aluminum spent fuel that involves melting the fuel along with depleted uranium in a plasma arc furnace at high temperature to produce a vitreous ceramic with a uranium-235 enrichment of 20 percent or less by mass.

Press and Dilute Treatment. A process for treating aluminum spent fuel that involves physically pressing the cut and sized spent fuel into sandwiches along with sheets of depleted uranium to produce dimensionally uniform packages with composite uranium-235 enrichments of 20 percent or less by mass.

Processing and Co-Disposal Treatment. A process for treating aluminum spent fuel that involves treating a portion of the aluminum spent fuel by conventional reprocessing and treating the remainder by direct co-disposal treatment.

Production Reactors. Nuclear reactors used to produce plutonium or tritium for weapons.

RBOF. Receiving Basin for Offsite Fuels.

Research Reactors. Nuclear reactors used for research and development activities.

ROD. Record of Decision.

SNF. Spent Nuclear Fuel.

TSS. Treatment, Storage, and Shipping Facility.

USNRC. U.S. Nuclear Regulatory Commission.

Vitrification. A process used to stabilize high-level waste that involves melting the waste in glass and solidifying the product in metal canisters.

WAC. Waste Acceptance Criteria, the physical, chemical, and thermal characteristics that spent fuel, high-level waste, and associated disposable canisters must conform to for disposal in a repository.

Waste-Package Performance Criteria. The physical, chemical, and thermal characteristics that a waste package containing spent fuel or its treated equivalent must meet to be acceptable for shipment to and emplacement in a repository container.

WSRC. Westinghouse Savannah River Company.